FRENCH HOTEL DESIGN
法国酒店设计

法国亦西文化 编 林明炘 译

辽宁科学技术出版社
沈阳

Contents
目 录

个性酒店

Conceptual and Creative Hotels

都会酒店
Contemporary Urban Hotels

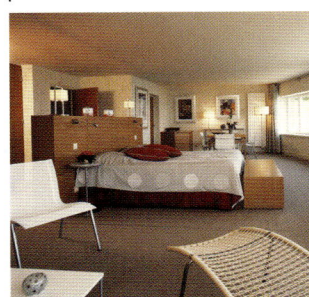

度假别馆
Dream Villas

附录
Annex

酒店资讯
Hotel
Directory

设计师资讯
Architect & Designer
Directory

Design - Innovation - Imagination

Conceptual and Creative Hotels

个性酒店

时尚–创新–概念

贝拉密酒店
Bel Ami Hotel

Designer:	Pascal Allaman (rooms), Nathalie Battesti & Véronique Terreaux (common spaces)	设计师:	帕斯卡–阿拉曼(客房), 娜塔莉–巴特斯提&薇洛妮卡–特若(公共空间)	
Location:	Paris, France	地点:	法国, 巴黎	
Completion Date:	2005	完工日期:	2005	
Photographer:	Patricia Parinejad	摄影师:	Patricia Parinejad	

Just a few steps from the church Saint-Germain-des-Prés, the Bel Ami Hotel revives the spirit of the avant-garde and offers a very designer address. Simple lines, open spaces, a playful approach to lighting, and natural materials in warm colours give the tone, in a relaxed and resolutely modern luxury environment.

Pushing open the door of the Bel Ami, visitors enter a world that is chic and convivial. The main body of the building shows off its original columns and windows, a memory of its industrial past. The original space has been completely restructured for a free circulation and larger volumes.

The hotel gives each room a different light according to its aspect and its floor. Some windows overlook the square of Saint-Germain-des-Prés. Rooms and suites combine modernity and warmth: the Orange Ambiance rooms display orange and grapefruit tones highlighted with spice-coloured lines, the Honey and Cumin rooms are light, contemporary and very elegant, while in the Pistachio rooms tender greens and cinnamon reign in soft tranquillity.

The B.A. Bar is decorated in a variety of warm shades of purple. Stained glass windows, wall and ceiling mounted lights create a play of light and vary the ambiance according to the hour of the day. Finally, the Bel Ami offers its guests a space dedicated to sport, relaxation and well-being: "Harmony" offers an oasis of peace at the heart of the city, under the benevolent gaze of fish of a thousand colours, ingeniously positioned in a wall-mounted aquarium.

位于圣杰尔曼–德培教堂附近的贝拉密酒店是一个风格前卫充满设计感的酒店, 其简洁的线条、开放的空间、变化的光线、自然的材料和温暖的色彩, 赋予这个华贵的酒店空间轻松的氛围与十足的现代感。

一进入酒店大门, 旅客马上可以感受到这里雅致而亲切的气氛。主建筑空间里保留着原始的柱列和窗户, 是对其过去作为工业建筑的一份纪念, 而经过重新安排组织后的室内环境则拥有更为顺畅的交通流线和更宽广的空间。

酒店的每一个房间因其方位和所在楼层的不同而具有各自独特的光线, 圣杰尔曼–德培广场的视野。所有的房间和套房都结合了现代感和温暖热情的氛围: 在称为"橙橘氛围"的系列房间里, 以加高彩度的柑橘色调为主色; 在称为"蜂蜜"和"豆蔻"的系列客房里, 呈现的是轻盈、高雅的现代感; 至于"黄连木果氛围"的客房, 则采用淡粉绿和肉桂绿, 让空间显得格外温柔宁静。

酒吧空间采用了带着温暖色调的红紫色系列, 其中的玻璃天顶、壁灯和天花悬挂灯件等, 各自提供了不同的光线组合与效果, 随着一日时间的变化而营造出多变化的氛围。此外, 贝拉密酒店也提供了一个能够让人放松身心、健身养神的"和谐"休憩中心, 其墙上水族箱里充满了五彩缤纷的鱼群, 整个氛围像是城市之中的一个和平的绿洲。

蒙马特府邸酒店
Hôtel Particulier - Montmartre

Designer:	Mathieu Paillard / Agent M Company
Location:	Paris, France
Completion Date:	2007
Photographer:	Marc Domage

设计师：	马修–拜亚 / Agent M公司
地点：	法国, 巴黎
完工日期：	2007
摄影师：	Marc Domage

Privacy is a true luxury. The choice of an hôtel particulier with only five suites, surrounded by a luxuriant and well cared garden, created by Louis Benech, offers the calm of the countryside in Paris. A group of artists has here created an interactivity between theirs and travellers' worlds.

On the ground floor, in the salons with windows outlined like paintings, Mats Aglund has installed first edition architect-designed furniture: the "Egg" armchair by Arne Jacobsen, the famous "Barcelona" by Mies van der Rohe or the creations of Le Corbusier... Each material and each object plays a coherent part in the score. You can read, have a drink or disappear in Olivier Sidet's mirror, anything is possible...The kitchen and the dining room, at garden level, symbolise the foundations of the house. In a baroque atmosphere, the furniture and objects hunted down at flea markets evoke the world of 19th-century grandmothers.

私密性在今日社会已成为一种真正的奢侈品！这个只有五组套房的府邸酒店，围绕着由景观设计师路易–贝涅克所营造的华丽花园，让旅客在巴黎市区里享受到乡间的宁静；艺术家们在这里将他们的宇宙呈现，希望能够与旅行者的世界交汇。

在地面层的沙龙空间里有着像是框景画框一样的窗户，芬兰设计师马特–哈古伦德在此设置了第一批由建筑师所设计的椅子：包括埃尔–雅各布森的蛋椅、密斯–凡德罗的巴塞隆纳椅、勒–柯布西耶的设计作品…等，每一个物件和材料都有其相互连贯性。在此，人们可以读书、小酌，或是选择消失在奥利维–西代所设计的镜子里，这里没有不可能发生的事。位于花园层、象征着家的厨房与餐厅有着巴洛克式的装饰风格，其中在四处旧货市场寻购而来的家具与物件在此营造出一种19世纪老奶奶世界里的韵味。

For the loft on the upper floor, Natacha Lesieur has painted eyes on the walls. In the centre of the room, two large portraits of women observe the visitors, their gaze hidden behind curtains of hair. Martine Aballéa wanted the guests in room n°4 to have the impression of finding themselves in a hanging garden or up in the trees ; of floating in an ethereal space, light, soft and flowery. The individual stands out by his absence in room n°3 by Olivier Saillard. A slender chair whose back finishes in a "butler" allows one to show off one's jacket. "Hat stand" bedside lamps make the bedroom into a dressing room where it is good to flop down like a shirt thrown over an armrest. Poems, like as many visiting cards that resident would be able to leave behind, decorate the walls.

In room n°2, a window, and a bedside table made of lights and mirrors, serve as the bed head. They reflect the coloured artworks of Philippe Mayaux. Finally, based on an Asian legend that recounts how secrets are confided into a hollow in a tree, room n°1 is, for Pierre Fichefeux, a place of confidences. The wallpaper, where indiscreet birds have been gagged, was created digitally. In the room, a tree has ears and records the words that are confided into, secrets revealed with no regard to propriety in the salon by the gossiping birds.

顶楼豪华套房的墙上有艺术家娜塔莎–蕾吉尔所画的眼睛，房间的中央有两幅女人肖像，其隐藏在发丝后面的双眼注视着来访的客人。玛婷–阿芭蕾雅希望让四号套房的旅客有一种身处在空中花园或是树林中的感觉，像是漂浮在一个居高临下、明亮轻柔而且繁花盛开的空间里。三号套房里奥里维–耶赛亚刻意让缺席的访客留下蛛丝马迹，一张纤细瘦长椅子的靠背形似衣架，让人可以吊挂西装；置帽架般的床头灯让人有仿若置身更衣室的错觉，忍不住要将自己当成是一件脱下来的衬衫一样，懒洋洋投掷在沙发椅背上，墙上装饰着一些像是以前房客所留下的诗句以及名片。

二号套房里有一座又是灯又是镜的床头镜板，其中映照着的是艺术家菲利普–马又的作品。最后，根据王家卫电影 "花样年华" 里要将秘密倾吐在树洞里的传奇故事，一号套房是由皮埃尔–费雪佛所塑造的秘密空间，利用电脑制作的壁纸画着被绑住嘴的鸟；在房间里，一棵有耳朵的树记录下人们对它的告白，而这些秘密，正被沙龙里多话的鸟儿们毫不害臊地讨论着。

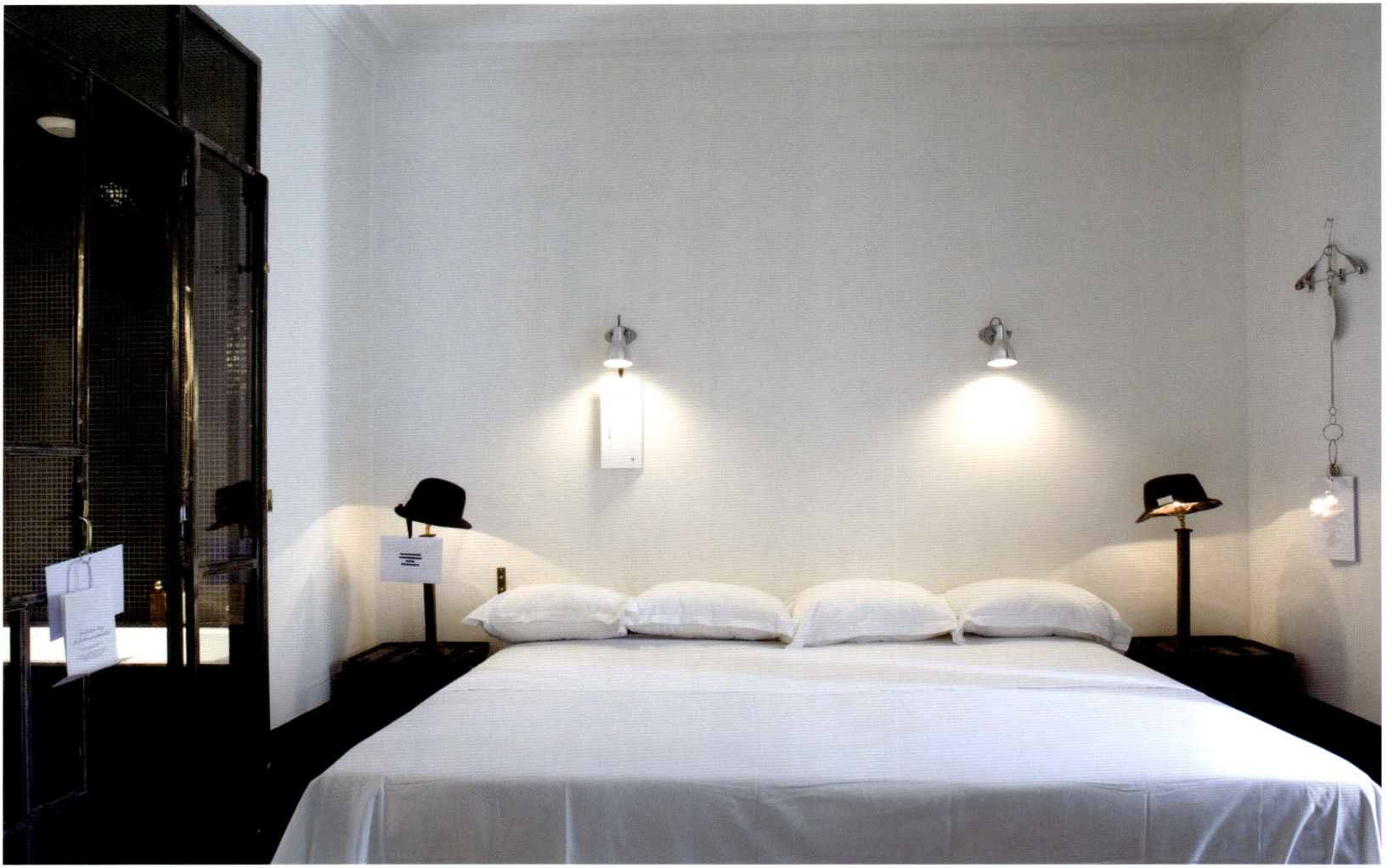

小磨坊酒店
Le Petit Moulin Hotel

Designer:	Christian Lacroix
Architect:	Vincent Bastie
Location:	Paris, France
Completion Date:	2005
Photographer:	Christophe Bielsa

设计师：	克里斯汀-拉克鲁瓦
建筑师：	文森特-巴斯提
地点：	法国，巴黎
完工日期：	2005
摄影师：	Christophe Bielsa

This hotel is made up of two old buildings joined together and refurbished by the Bastie architectural partnership. Not so long ago, the first building housed a bakery. And not just any bakery: the oldest one in Paris, dating from Henri IV. The local legend is that Victor Hugo used to come and buy his bread there. The 1900 frontage is listed by Historic Monuments as part of French heritage. The interior decoration of the old bakery, going right up to the ceiling, has been restored and creates a "Venetian" style setting for the reception area.

Christian Lacroix, who designed the hotel, immediately liked the slightly crooked and off-kilter perspectives. The new, very functional spaces, have been created while respecting the picturesque "old Paris" look of the listed elements, like the lovely 17th-century wooden staircase that has been left in its natural state. All of these very special volumes dictated one single idea that seemed to him to be coherent with such a place and location: each room really had to be very different, in relation to its orientation, the height of its ceiling and its location within the heart of the hotel, and even the view. But most importantly, each room had to relate the beginning of a story to be finished by the guests themselves. Anne Peyroux has helped him to realise this utopia.

Beams or cement, antique wallpaper or gaudy fabrics, contrast with wooden floors and tiles, carpeting and ceramics. An entire collection of contemporary lighting units, of 60s seating covered in brocade or velvet, or a little bit of faux fur and some period furniture upholstered in very bright colours and graphics, constitute the vital thread linking this medley of different worlds and finishes. The tone for this patchwork of moods has been set via panels of giant collages on the walls or windows.

位于两栋旧建筑物内的小磨坊酒店，由巴斯提建筑师事务所将其连接并重新整建。第一栋旧建筑在亨利四世的年代曾经是巴黎第一家面包店的所在地，传说大文豪雨果也到此地来购买面包。其建造于1900年代的店面被法国政府指定为历史性遗产而保存，店面里一直延伸到天花板的、覆有玻璃的小油画装饰也被精心修整，使得酒店接待空间呈现出像是威尼斯一样的璀璨风华。

为酒店进行室内设计的克里斯订-拉克鲁尔立刻爱上了酒店里不规则的空间透视效果，这些新建完成、非常具功能性的空间，却也善用被列为保护的建筑元素而塑造了非常具老巴黎味道的、秀丽如画的风格，例如17世纪的美丽木造楼梯。在这个各自具有独特体量的空间中，设计师认为每一个房间必然得呈现完全不同的面貌，依照它们各自的方位、空间高度、在酒店内的位置、甚至景观视野等等不同的条件，来赋予每一个房间特殊的性格，并营造一段故事的开端，让在此停留的旅客能够继续延续故事、留下一点个人的蛛丝马迹。他在安-培鲁的协助下将这个乌托邦式的构思实现了出来。

木梁或是混凝土、壁纸或是有花边装饰的织品，搭配着木地板与陶土地砖、地毯与瓷砖。一系列的现代灯具照明、以小羊皮或是绒布覆面的座椅、还有毛皮和带有鲜艳色与彩图案的家具，构成了一道不可或缺的设计线条，来连结这个具有混搭性格的世界，配合着隔墙和窗户上的大幅拼贴作品，赋予这里一种犹如碎布拼缝成的百衲被氛围。

The decor plays on contrasts: from lawn green corridors with black lacquered doors framed in optical white on a polka dot carpet background to decors designed with tongue-in-cheek pilasters, cornices and consoles ; from modern bathrooms in slate, faïence or cement given warmth by Venetian mirrors, ceramic kaleidoscopes or panoramic wallpapers to baroque, rococo or "Couture" rooms ; and from 21st-century spareness to a wink at nostalgia. The bar, treated like a local street café, combines the zinc counter and "Edwardian" furniture with "scrap-book" walls and 60s seating in shades of yellow and pink.

人们从衬着小点状地毯与白框黑门的绿色英国式廊道，来到带有壁柱、墙缘突饰和托架等建筑元素、令人莞尔一笑的空间装饰；从运用了黑色板岩、马赛克、混凝土来装修，并带有威尼斯镜面、万花筒式瓷砖以及大面花色壁纸的浴室，来到具有巴洛克和洛可可风格的空间；从21世纪的纯净极简，来到怀旧气息。设计得像是街区咖啡店的酒吧空间，将锌质材料与巴黎"美好年代"风格的家具相结合，并搭配着红黄双色的60年代座椅以及"剪贴簿"似的墙面。

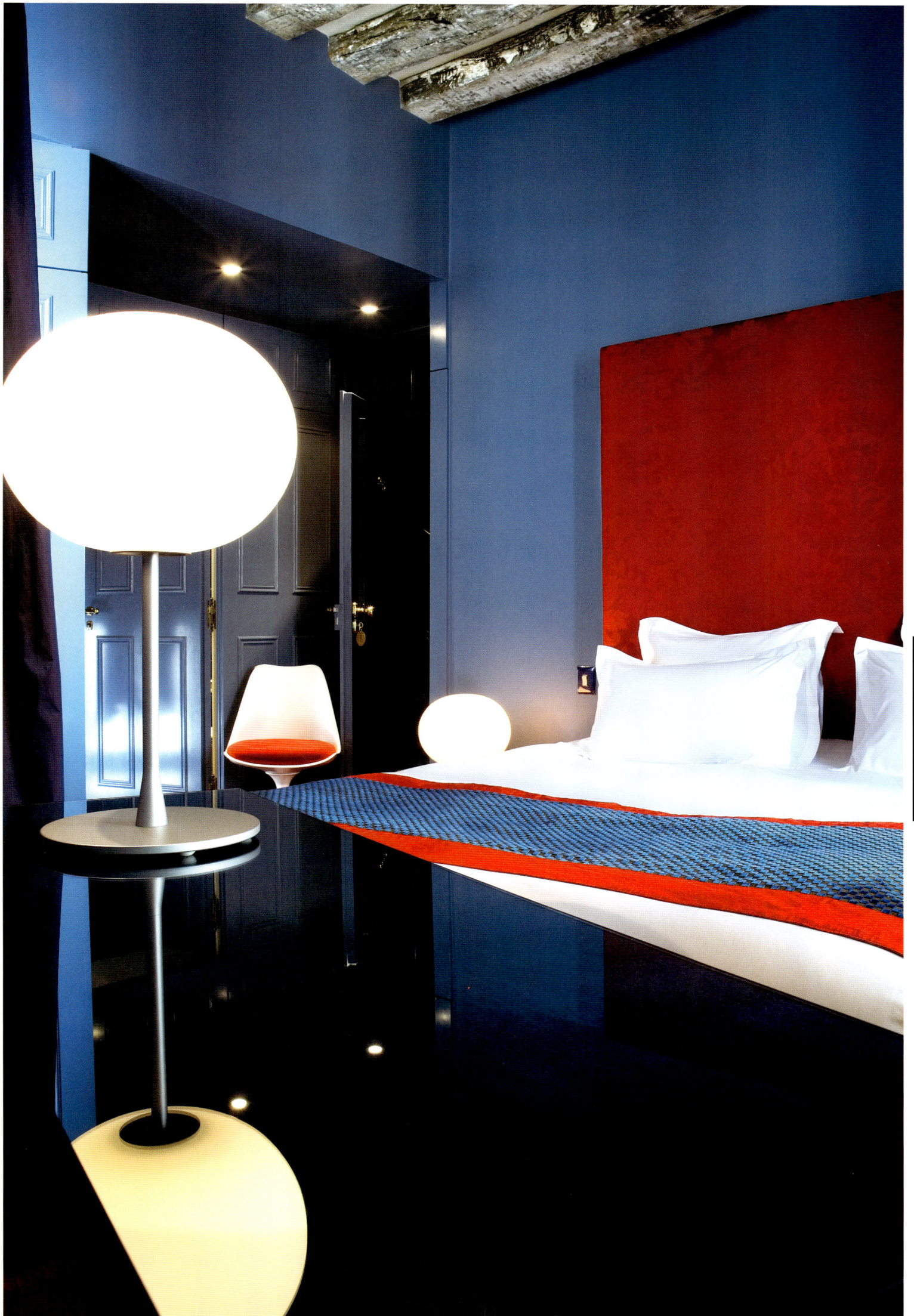

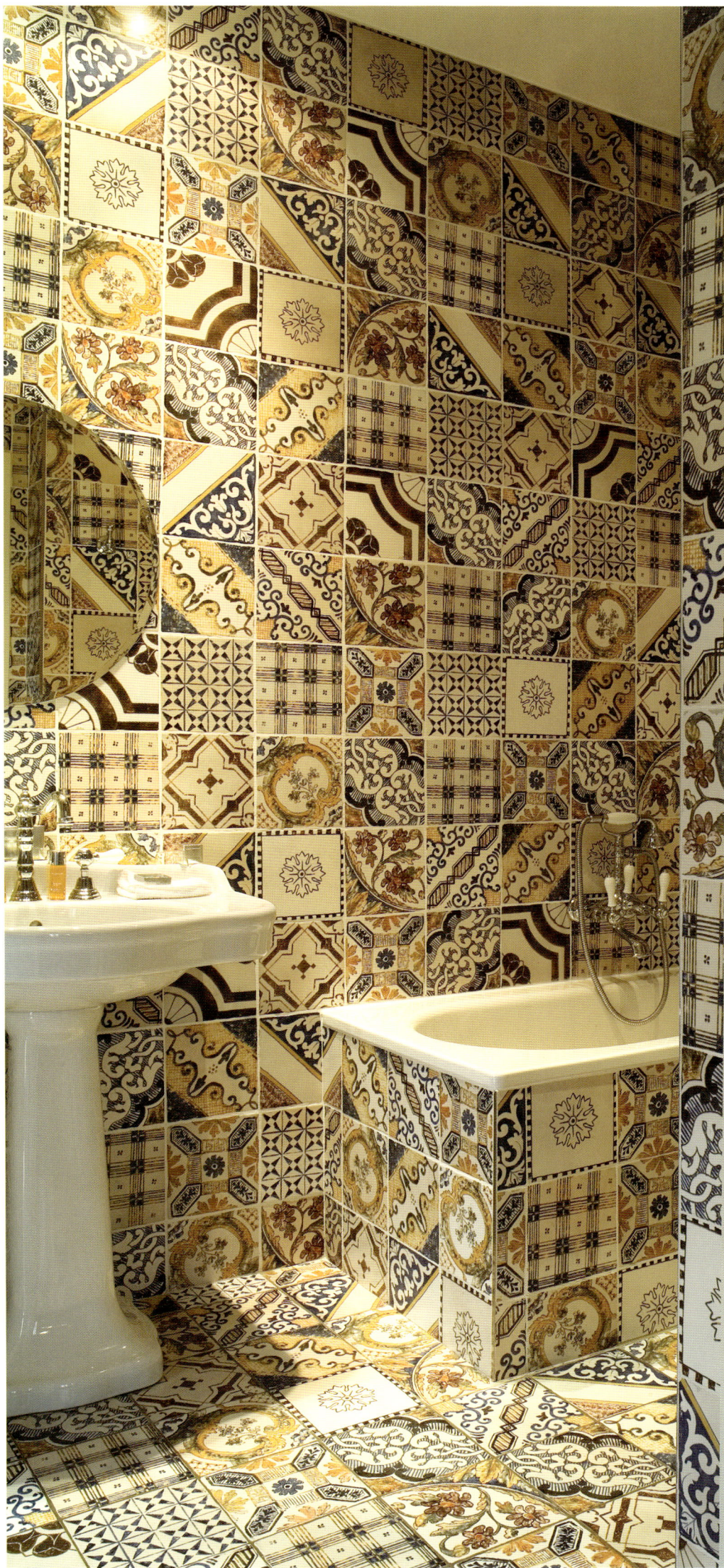

雷克斯酒店
Le Rex Hotel

Architect:	Joris Ducastaing
Designer:	Isabelle Bonis
Location:	Tarbes, France
Completion Date:	2006
Photographer:	Ralph Hutching

建筑师：	裘里–达卡斯堂
室内设计师：	伊莎贝拉–伯尼斯
地点：	法国，塔贝斯
完工日期：	2006
摄影师：	Ralph Hutching

In the heart of Tarbes, enlivened by the changing colours of a huge glass facade, this resolutely design establishment never fails to delight its guests. An edifice created from glass, concrete and steel, its impression of lightness provides endless fascination.

The very contemporary design of the hotel, which nevertheless references the traditional architecture of Tarbes, was entrusted to Joris Ducastaing and Jean-Luc Heins. Everything started with the superb facade covered with an opalescent glass skin, which is uplit by fibre optics at night to look like a kaleidoscope. The central hall reminds one of a cathedral with a huge light shaft rising through five floors and, as a focal point, a huge chandelier in transparent PVC tubes that changes colour over time. As for the furniture, here you'll find the great classics of the Scandinavian and Italian designers.

The 92 rooms, including five terraced suites on the 7th floor, have an intimate atmosphere with walls and carpets in a chocolate colour, setting a scene for a contemporary space initiated by a strong architectural choice: a bathroom like an aquarium made up of glass panels, where only a play of filmy fabric assures privacy. Here, too, modern and pure design is the chosen option.

The RexLoundge, a bar enclosed by a barrier of white coral and crowned with an aquatic glass ceiling, is an island of glass and of light. The RexCafé is a more minimalist restaurant, black and white but just as welcoming, and the RexMeet serves as a meeting place for business guests. The "Athena" and "Pandora" rooms offer privileged meeting spaces equipped with state-of-the-art audiovisual equipment.

雷克斯酒店地处塔贝斯市的中心，其晶莹多彩的立面为酒店外貌带来极强的韵律感，整个酒店充满设计感，所有构思皆为了满足旅客的各种需求。建筑物运用了玻璃、光面混凝土、钢和不锈钢等材料建造，散发出轻盈迷人的魅力。

这个带有塔贝斯地区传统建筑特色却又极具现代感的酒店，是设计师裘里–达卡斯堂以及让吕克–汉斯的杰作。建筑物的立面覆盖着一层透光玻璃表皮，在夜间向外放射出犹如万花筒式的光芒。酒店的接待大厅拥有一个4层楼高的采光井，令人感觉像是一座大教堂，其中巨大的PVC透明光柱，随时间而变换颜色，仿佛空间的延长号。在家具方面，则主要采用了北欧以及意大利设计师的经典作品。

酒店总计有92间客房，其中5间配备了观景露台的豪华套房分布在8楼。在温柔舒适的空间气氛以及墙面和地毯中的巧克力色调当中，每个房间都特别设置了一个现代感十足的建筑元素：一个由玻璃筑成、有如水族箱一般的浴室。一袭薄纱帘幕便解决了私密性的问题，设计师在此追求的是一种纯净现代的空间风格。

雷克斯酒吧由周围的白珊瑚护栏以及顶上的玻璃采光罩来界定空间，仿佛一个明亮通透的独立区块。雷克斯咖啡是一个以黑白双色为主的极限主义餐厅，简洁中却不失亲切迎人的气氛。雷克斯会晤区则是专门提供商业人士约会议事的空间。雅典娜沙龙与潘朵拉沙龙是保留给特殊场合所使用的贵宾空间，具有最完善的视听设备服务。

皮尔辛府堂酒店
Pershing Hall Hotel

Designer:	Andrée Putman, Imaad Rahmouni	设计师：	安德蕾–普特曼, 伊玛德–哈姆尼
Location:	Paris, France	地点：	法国, 巴黎
Completion Date:	2001	完工日期：	2001
Photographer:	Ilan Assayag	摄影师：	Ilan Assayag

Located in the heart of the Golden Triangle, this classy hotel contains 26 rooms, with 6 suites and 4 junior suites, a restaurant, a lounge bar, reception rooms and a spa. The facade and the wide flight of stairs have been here since the Count of Paris had this mansion house built. When the United States declared war on Germany in 1917, the building became the headquarters of Commander in Chief John J. Pershing.

The most talented architects of the present day have now given its rooms their meaning and rhythm. Andrée Putman brought soul and rigorous grace to the decoration of the various rooms. She designed the furniture for every room. Imaad Rahmouni brought the textures of a dream Mediterranean: Murano crystal, strings of pearls and silk cushions create a soft and delicate atmosphere in the lounge and the restaurant.

Every room is charming and elegant in a refined and luxurious style, with a subtle play of light and harmony in the colours and materials. Each room is organised around a large three-pane mirror and a high wooden bed lacquered in ivory. Every detail is subtly arranged, from the lighting to the arrangement of the three-patterned curtains reflecting the Paris sky. The bathrooms have been lined in glass mosaics in opalescent white and ivory, enhanced by golden accessories. Bathtubs are set on white marble feet. Works of art by contemporary artists add the final touch with a mixture of bright and deep colours.

位于巴黎金三角地带、优雅的皮尔辛府堂酒店共有26间客房，其中包括10间豪华套房。此外还有餐厅、交谊吧台、接待处以及养生水疗中心(spa)。酒店的建筑原是由巴黎伯爵所建造的私人宅第，它的立面以及内部的主要楼梯至今都还保存着原貌，其他的部分则因着时代的变迁而有数次的更动。在1917年，一次世界大战期间，以约翰–皮尔辛将军为首的美军曾经在此处驻扎，该建筑因而成为当时的指挥所。

在当今法国最具才华的建筑设计师之一安德蕾–普特曼的巧意构思下，酒店的空间与其装饰风格得以重新获得秩序、具有新的意涵与韵律节奏。她为酒店的整体空间组织带来了灵气与庄重审慎的雅致感，还亲身设计了酒店里的每一个房间的家具。另一位设计师伊玛德–哈姆尼将引人梦想的地中海精神带到酒店当中，意大利的穆拉诺水晶、丝织的绒球以及靠垫，让人身在酒店的餐厅和交谊厅时倍感轻松舒适，享有一种异国风情。

酒店的客房散发着一种优雅柔美的氛围，带有净简的华丽、细致的光影变化，以及色彩与高贵材质间的巧妙结合。在每一个房间里都有一面三折式的大镜和一张像船一般的象牙色木制高床；在此气氛与光线下，一切都显得非常雅致，甚至三个如画作般框着巴黎天空的窗景与窗边大纱帘的结合亦不在话下。客房浴室的墙面装裱着玻璃马赛克：一种半透明白色或是象牙白的石材，被缀入其中的金色宝石衬托得更加亮眼；此外，高高坐在四颗大理石石球之上的白色浴缸，亦是设计的妙笔之一。

The living areas are set around a patio opening up to the sky. An incredible garden rises vertically through more than 30 metres. This wild but carefully tended garden is composed of 300 different species of tree, shrub and plant from the Philippines, the Himalayas and Brazil. Creepers, ferns, giant-leaved plants and orchids growing overhead give a dizzying sensation. This horticultural triumph is the work of the famous botanist and garden designer Patrick Blanc.

Visitors arriving for the restaurant and the lounge pass through the entrance hall with its long curtains of glass beads. Lamps and vases from crimson blown Murano glass create a dialogue with the tamed jungle of the vertical garden, which dominates the courtyard. Red, green, golds, all the tones of controlled passion... In the courtyard restaurant neighbouring tables strike up conversation or compete with each other. The lounge is reached by the wide, star-spangled staircase. Its balconies open over the courtyard to provide viewpoints on the restaurant, the sky and the plants. The feeling of relaxation is confirmed in the huge, comfortable sofas and armchairs in beige leather with silk cushions thrown over them.

在酒店建筑物素朴的立面后方，其公共空间围绕着内部露天的中庭而展开，此中庭出奇之处在于它沿着墙面爬升高过30公尺的垂直型花园，这个充满自然野性、却受到悉心照顾的花园有着来自菲律宾、喜马拉雅山区、亚马逊河流域等地超过300种以上的树木和灌丛，各式藤蔓、蕨类、肥厚硕大的树叶、还有一支超乎想象异常高大的兰花，这个园景奇观要归功于知名的植物学家与园艺师派崔克–布朗。

缀有大幅玻璃珍珠球的帘幕带领人从大厅前往交谊酒吧与餐厅空间。这里有着由意大利穆拉诺玻璃吹制成的花瓶与灯饰，大红色的镜子对应着外面中庭主墙上象是被驯服的绿色丛林；空间中使用了红色、绿色、金色…等强烈而不夸张的色彩。在中庭里或嘈嘈细语或相互排挤的几张小桌属于餐厅空间，通过带有星形扶手的宽广楼梯得以前往酒吧交谊厅。交谊厅的阳台拥有面对餐厅、天空和绿色植栽的多变视野，人们坐在羊毛灰色、带着丝织靠垫的舒适大沙发上，得以尽情享受此地的轻松氛围。

法福酒店
The Five Hotel

Designer:	Vincent Bastie	设计师：	文森特-巴斯提
Location:	Paris, France	地点：	法国,巴黎
Completion Date:	2006	完工日期：	2006
Photographer:	Christophe Bielsa	摄影师：	Christophe Bielsa

In the Latin Quarter, which since the 1920s has been a centre of artistic and student life, The Five Hotel is ideally placed to discover and make the most of Paris.

From its very conception, The Five Hotel prepared to welcome guests in a warm designer setting that charms the five senses. Smell, thanks to rooms delicately perfumed with a choice of five aromas ; touch, by the presence of fine and original materials: carpets, velvet, sandstone... at The Five, design is made cosy. The sense of hearing is perfectly rested thanks to the calm of the environment, reinforced by double glazing. Sight can rejoice in an ultra-design decoration where light from fibre optics is employed to give each room a psychedelic allure, with desks of luminous glass, stars on the ceilings, and light sculptures for a magical glow. As for taste, that comes in the form of a crunchy breakfast taken among lively colours which perfectly energise The Five Hotel.

The ultra modern works that one can find from the reception to the rooms also encourage you to dream, to feel that you are somewhere else. They are the work of Isabelle Emmerique, an internationally recognised artist who has been working with Chinese lacquer for the past 30 years.

巴黎的拉丁区从1920年代以来就以其充满艺术与学术气息闻名而成为重要的观光地点，而位于本区的法福酒店正是可以让旅客尽情地探索和享受巴黎的下榻之地。

从接待处开始，旅馆就以其充满热情和设计感的方式来表达欢迎之情、提供旅客在五种感官上的享受。客房里设置了可任由房客选择的五种不同香气来引起嗅觉的愉悦；运用高贵且具原创性的材质，像是织毯、天鹅绒以及陶土砖…等，来塑造细微、特殊的触感；由于地理环境本身宁静特性，加上双层玻璃窗户的加强，使人们在室内的听觉得以获得良好的休息；在装修上极富设计感的客房里，运用光学纤维塑造出的光线效果，加上玻璃制的书桌、仿佛夜空繁星闪烁的天花板以及雕塑造型灯具的神奇照明，使得每一个房间都具有不同的视觉幻象；此外，酒店所提供的香脆早餐是为味觉所作的精心设计，在鲜艳色彩当中享用它，更传达了酒店充满活力的意象。

从接待处到房间的种种具有超现代感的艺术作品带领着旅客进入一个令人向往的境地，仿佛身处另一个世界，这都要归功于国际知名艺术家伊莎贝尔-艾美里克30年来对中国漆艺的研究与创作成果。

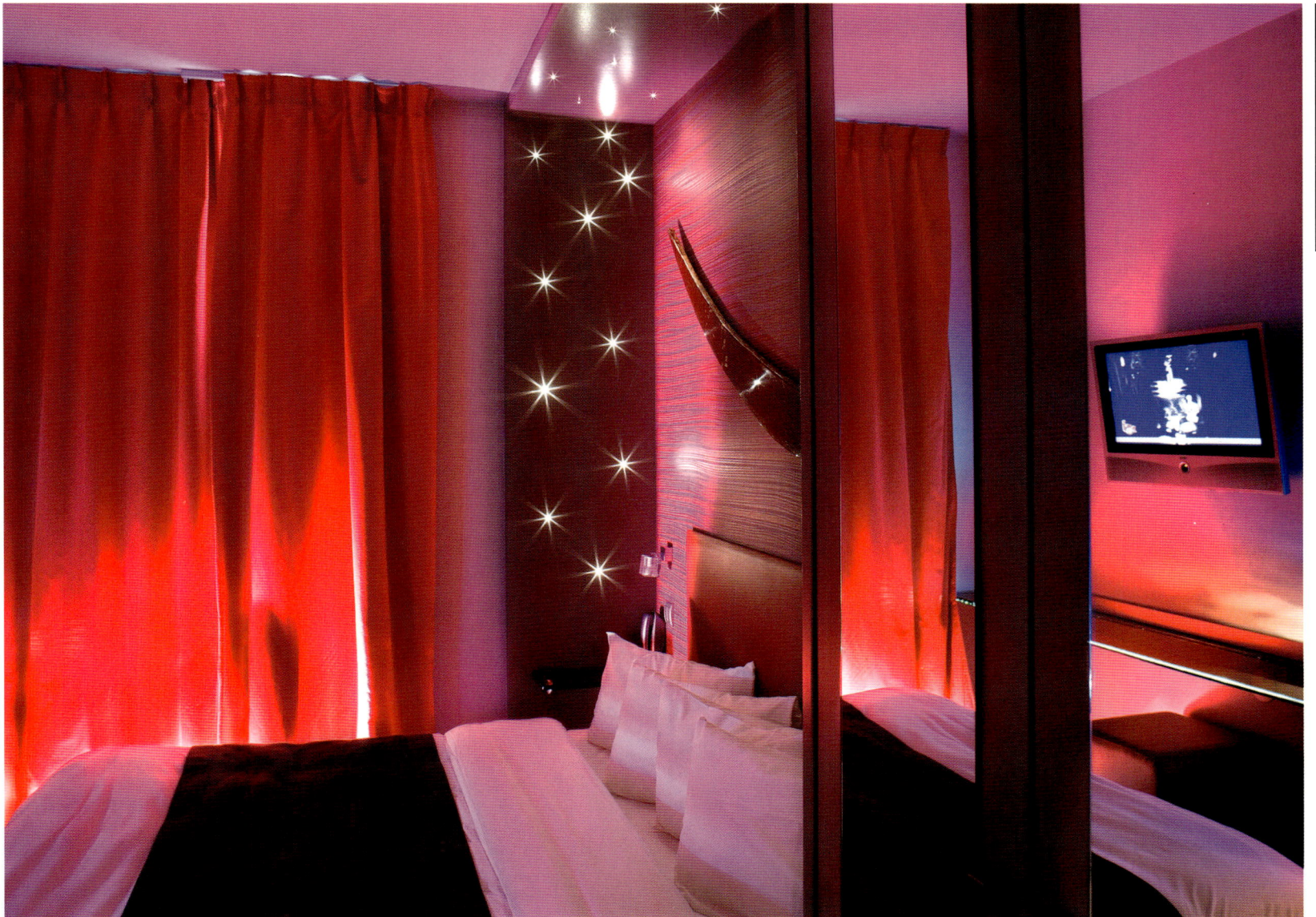

贝乐夏思酒店
Bellechasse Hotel

Designer:	Christian Lacroix	设计师:	克里斯汀-拉克鲁瓦
Architecte:	Jean-Luc Bras, Emmanuelle Thisy, Anne Brugière Peyroux	建筑师:	文森特-巴斯提, 埃曼纽尔-提斯, 安布鲁姬-裴忽
Location:	Paris, France	地点:	法国, 巴黎
Completion Date:	2007	完工日期:	2007
Photographer:	Christophe Bielsa	摄影师:	Christophe Bielsa

Close to the Musée d'Orsay, the Bellechasse is a delightful hotel, and full of character. "Dressed" by Christian Lacroix, it manages to be both impressive and intimate. Behind its classic appearance is hidden an unusual personality: 34 rooms spread over two buildings, where designs, figures, colours and subjects combine the inspirations of neo-classicism and the bohemian spirit.

Playing with a palette of ambiances and influences, perpetuating a skilful balance between audacity and tradition, Christian Lacroix created seven themes that are deployed through the various rooms: "Patchwork", with Persian and Arabic motives, "Avengers", with a hint of the 1960s television series, "Saint-Germain" in a contemporary style, "Tuileries", folk motifs in black and white, "Mousquetaires", in yellow brocade, rough stones and touches of velvet, "Jeu de Paume", neo-futurist and abstract with a base of lively and primary colours, and, last but not least, "Quai d'Orsay" with murals recalling the world of the poet and screenwriter Prévert, or with astrological motifs.

For Christian Lacroix a hotel must reflect the character of the locality it is standing in and should represent "travel within travel" while giving its own interpretation of the city, the district, the street it opens on to. The hotel Bellechasse is the result of this skilful alchemy. The baroque inspiration of the fashion designer is mixed with the influences of a highly historic district. A mixture of shapes and colours makes it most attractive.

临近巴黎奥赛美术馆的贝乐夏思酒店请来了时尚大师克里斯汀–拉克鲁瓦为其进行室内设计, 成为一个同时具有声誉与私密氛围的魅力酒店。在其建筑的古典立面背后, 隐藏着异于常规的特殊性格, 其分布在两栋建筑物之中的34个客房, 各自以不同的图案、影像、色彩与材质, 展现出融合了新古典风格与吉普赛精神的设计风貌。

克里斯汀–拉克鲁瓦运用一系列的氛围与效应, 来持续展现一种介于传统与果敢作风之间的微妙平衡, 在所有客房当中塑造出7种不同的世界: "百衲拼组"带着阿拉伯风格以及波斯图案中的原始色调; "复仇者"是对英国电视影集"哈密瓜帽"与"皮靴"的致意; "圣杰尔曼"有着极为当代感的设计风格; "图勒里"在白色底衬上挥洒具有乡村风味的黑色花纹; "三剑客"的粗石搭配着黄色锦缎与丝绒; "网球场"以鲜艳的原始色调展现着新未来派与抽象派风格; 最后的"奥赛堤岸"不是有着星象学壁画, 便是带着雅克–普莱维尔诗句意象的墙面。

对克里斯汀–拉克鲁瓦而言, 一个酒店的设计必须反映出所在地点的特色, 并提供一种"在旅行中旅行"的感觉, 以独特的方式诠释城市、街区与街道, 而贝乐夏思酒店就是这种精致提炼的成果。在一个充满历史痕迹的街区加入时装设计师的巴洛克式灵感, 酒店的魅力便来自于这种洋溢着愉悦感的形体与色彩的错综交合。

三点一四酒店
3.14 HOTEL

Designer:	Karine Ellena-Partouche, Alexandra Ellena
Location:	Cannes, France
Completion Date:	2004
Photographer:	Morgan Rouillon (pp.60-61), Alban Pichon (pp.62-64), Marcel Partouche-Sebban (pp.65-67)

设计师：	卡琳娜–艾莲娜–帕尔度叙，亚历山德拉–艾莲娜
地点：	法国，戛纳
完工日期：	2004
摄影师：	Morgan Rouillon (pp.60-61), Alban Pichon (pp.62-64), Marcel Partouche-Sebban (pp.65-67)

The 3.14 is pushing back the boundaries. Open to other civilisations, the designers have always dreamed of uniting in a single place the atmospheres that have impregnated and formed their two personalities: the scents, colours, materials, music and tastes of different countries in the world...

Each floor represents one of the five continents in a scene created according to the principles of Feng Shui. The rooms on the first floor plunge visitors into a multicultural America crossbred with Latino culture. The kitsch Pop Art decor emphasises the eclecticism of the New World's melting-pot... Then, the Africa of the 1001 Nights. From the Moorish windows of the second floor rooms, billowing curtains decorated with beads evoke the sensuality of a belly dancer. How could one fail to succumb to the charms of the distant islands of the South Pacific? The harmonious lines of the furniture, the symbolism of the decorative motifs, the raffia shades... are all reminiscent of the original paradise.

On the fourth floor, Paris of the Belle Époque composes a voluptuous symphony where the pink wall drapes and padded bedheads mix with crimson red curtains. The soft wall lights reveal the femininity and romanticism of this place. And finally Asia, where, lightly scented by aromatic incense, the rooms are inspired by the traditions of Japan, and an ever-present Zen procures a feeling of deep serenity. The geometric lines of the dark furniture combined with the lighter shades of the natural fabrics and the brightness of red Chinese silk lead one's body and mind towards peace. Buddha watches over the guests as they sleep.

On the roof top, between a bamboo forest and a waterfall, the swimming pool and the Jacuzzi face the Mediterranean sea. Massage, energetic relaxation, a hammam and fitness centre are designed to ensure harmony under a cerulean sky.

三点一四酒店的两位创办者与设计者希望透过这个场所为人们拓展视野疆界，希望能将增经陶冶与塑造她们性格的各种环境氛围全部集中在一个场所里，融合一炉而治：气味、颜色、材质、音乐以及世界各地的美味……。

酒店的每一层楼都融合了风水的观念来设计，并且分别代表了世界五大洲的不同特色。首先，位于二楼的客房使旅客沉浸在混合着拉丁美洲风格的空间氛围，樱桃酒红色的波普艺术装饰品让人感受到犹如民族大熔炉的新世界性格……；接下来的三楼代表着非洲一千零一夜的传奇，有着摩尔人建筑风格的窗户，搭配着令人联想到中东舞者的大幅珍珠薄纱帘幕；此后，轮到遥远的大洋洲诸岛在此展露它们的风情：家具的和谐造型、装饰图案的象征性格，加上用椰叶纤维制成的百叶窗帘，呈现了大自然的所有奥秘。

五楼的空间塑造出由巴黎美好年代所代表的欧洲风格：墙面的装饰色调、床头软垫的玫瑰粉彩与窗帘的胭脂艳红相互搭配融混，像是一曲悦耳的交响曲；壁灯射出温和的灯光让整个空间充满了一种女性柔美的浪漫特质。最高楼层的空间以亚洲作为主题，客房装潢采取了传统的日本风格，并加上一些综合性的清淡芳香，塑造出无处不在的禅意，并带给人一份宁静泰然的深刻感受；此外，暗色家具的几何线条搭配着自然布料的明亮以及中国大红丝绸的亮丽，将人领往一种绝对和平的境界。一旁静坐的佛像守护着旅人的安眠。

屋顶平台上有着完善的健身设备：位于竹丛与小瀑布之间的游泳池与按摩池享受着面对地中海的景观；而按摩室、能量放松室、蒸汽浴室和健身中心等空间的设计，让人仿佛徜徉在蔚蓝天空下的和谐气氛中。

皮克之家
Maison Pic

Designer:	Philippe Puvieux		设计师：	菲立浦-普尤
Location:	Valence, France		地点：	法国, 瓦朗斯
Completion Date:	2008		完工日期：	2008
Photographer:	Gil Lebois (pp.68-69, 70 bottom right, 71, 73-75),		摄影师：	Gil Lebois (pp.68-69, 70右下图, 71, 73-75),
	Jean-Yves Salabaj (p.70 bottom left), Maison Pic (p.72)			Jean-Yves Salabaj (p.70 左下图), Maison Pic (p.72)

The hotel is luxurious, calm and soothing–but with a spirit that's contemporary, chic and occasionally impertinent. Absolute comfort and style come together. Here, people are not at home but in a hotel, La Maison Pic, and thus are seeking spaces that don't remind them of home but are different, unexpected and inviting.

Behind the patio's arcades, a banquette stretches out indefinitely, lined with small tables made of dark wood. In the living rooms giant sofas face each other–Alice would have loved them, for they would look at home in Wonderland. They resurrect the pleasure of languid conversation. Everything here invites one to spend time as a couple or with friends, talking, exchanging ideas and reinventing the world as people sink into the elegant sofas. Graphic bunches of flowers, improbably shaped branches–style is everywhere.

In the restaurant Anne-Sophie Pic, as in an old bourgeois house, successive rooms lit by French windows open onto the patio or the adjacent gardens. Bright colours and period furniture provide unexpected notes thanks to their colour and size. Enormous frames without pictures, tables dressed with bursts of light and simplicity, the gentle murmur of conversation, and the discreet presence of the waiters and sommeliers all add to the atmosphere.

豪华、静谧、令人迷醉的特质，再加上灵性、现代、雅致以及偶尔的放肆，在提供极致舒适的客房的前提下，皮克之家想要塑造的并非旅客们自己家的氛围，酒店主人认为来到这里的旅客并不想寻求复制与他们在别处有过的相似生活场所，而是希望经历不同的空间体验，就像一场真正的旅行、一个短暂的轴线偏离，一个特殊的邀请。

在中庭回廊之后，靠墙展开了一组软垫长凳，搭配深色原木独脚圆桌……，沙龙区中则以面对面的方式摆放着大型软垫座椅，让人重拾对话的乐趣。这个精心设计、重新诠释的交谊空间，让人们可以将自己舒适地安置在优雅的沙发中，尽情享受二人交心或是与众友畅谈的乐趣。此外，精致的插花艺术，使得空间到处充满个性。

就像在过去大户人家的宅第里一样，酒店里的安苏菲餐厅是由一连串的小型餐间所组成，一旁的大片玻璃窗带来明亮光线，并使餐厅得以享受酒店内庭花园的景致。鲜艳色彩与古老家具的搭配，使得空间的色调与体量感产生错位与颠覆。映照着光线的简洁餐桌、客人的低声对话，以及服务生与品酒师低调的往返服务，构成了餐厅的各种画面。

The only place that looks onto the street, but also the patio, is Le 7. Mirroring the Nationale 7, the route of yesterday's gastronomads, this bistro takes all the symbols of the road and reinterprets them in a decor that is baroque and full of humour. Napoleon III chandeliers, white dishes and red glasses, road signs on the floor, milestones, a bar for quick meals, Plexiglas chairs, bistro tables with knobs for hanging your napkin, everything in red, asphalt grey and white under an open sky and in the shade of plane trees photographed in black and white–the atmosphere is lively and chic! A folded map serves as a menu.

The hotel's rooms, in a spirit both contemporary and chic, are luxurious and calm. With no frontier between bedroom, bathroom and living room, they set the scene for languorous conversations. With a small salon, a walk-in wardrobe, a huge shower as well as a bath, a flat screen for watching films in dizzying clarity, and a patio terrace for each room, nothing has been overlooked. There are also very comfortable rooms in a more authentically Provençal style, with distressed furniture and warmly welcoming bathrooms.

七号酒馆是唯一一直接面对户外与建筑物中庭的空间。七号，就像国道七号是过去著名的美食之路一样。这是一个酒馆餐厅，它以巴洛克式风格加上满心的幽默，重新在此诠释了国道七号美食之路上的精华意象：拿破仑三世的玻璃吊灯、白色餐盘与红色酒杯、地板上的图案，让人快速品尝美食的吧台、有机玻璃制的座椅、设有铜扣让人便于固定餐巾的餐桌。这一切都在红、灰、白三色之间调和，上方顶着天蓝以及印着梧桐黑白影像的天花板。

在客房的设计当中，设计师企图在彼此相通的卧室、浴室以及起居室之间塑造一种断续不连贯、却彼此对话的空间。在每一个面对中庭的房间，睡床、浴缸、小起居室、更衣室、宽敞的淋浴间、放映着惊心动魄影片的超薄大荧幕以及私人露台，都参与了这里别出心裁的设计。酒店里另外有一组具有普罗旺斯典型风格的舒适客房，在充满阳光色彩的空间下和温暖热情的浴室中，搭配着充满历史感的古式家具。

新马赛酒店

New Hotel of Marseille

Designer:	Alain Sarles / Archimed	设计师：	亚兰–萨尔勒 / 阿基米德公司
Location:	Marseille, France	地点：	法国，马赛
Completion Date:	2006	完工日期：	2006
Photographer:	Erghot Groupe New Hotel	摄影师：	Erghot Groupe New Hotel

The New Hotel of Marseille has opted for a resolutely eclectic style. The entrance, facing the Palais du Pharo, is a listed 19th-century monument, the old Institut Pasteur. Beautifully renovated, it plays with sobriety and modernity, with charcoal grey floors and white walls. Metallic 1950s loft-style lights illuminate a black counter. A pale green armchair enlivens the scene. The reception is linked to the main building by pared down architecture and a rectangular glass-roofed patio.

Surprises are everywhere: 19th-century armchairs with ear pieces contrast with a Murano chandelier under the glass roof, with contemporary artworks in the mix. Everything is a pretext for a work of art. At the top of the majestic staircase is an impressive canvas typical of the work of Tomas L. Jacques Combet, alias Cabane, prepares the ground for the floors above with a symbolic arrangement of cubes placed side by side, and, en route for the rooms, the corridors are rainbow-like with their play of pink, yellow, blue and green lights.

On the upper level, the spacious central bar is reminiscent of the 60s with a flourescent green theme, a charcoal grey floor, dark wood tables and grey and red chairs. To the side, tables and chairs in wenge wood mounted on slate slabs furnish the restaurant. Sobriety comes to the fore in the rooms. Grey-black tones, a flat screen TV and dark wood furniture with a few touches of sober colour create the image of the hotel. The walls are hung with fabrics printed with black and white photos, evoking fashionable travel images of 50s.

Finally, the teak terrace, situated just under the Fort Saint-Nicolas, offers a pretty view of the Vieux Port. The pool situated here has, since 2007, been adorned by a 100-foot vine of Syrah grapes.

新马赛酒店是一个融合众多风格的地方。酒店的入口是昔日巴斯德研究中心所在地，面对法罗宫殿，是被法国政府指定为历史文化资产来保存的19世纪建筑。此空间如今经过细心的整修后，展现了简洁而深具现代感的风格：深灰色石造地板衬着雪白墙面，50年代的金属大吊灯映照着其下黑色的吧台，空间中淡绿色的沙发因此显得格外亮眼。此入口空间通过罩着玻璃天顶的长方形中庭通往主要酒店的建筑。

酒店里到处可见的是19世纪风格、带着护耳边的扶手椅，玻璃天顶下吊挂着一盏意大利穆拉诺的玻璃吊灯，此外还有众多当代工艺作品。这里不乏艺术展现身手的机会：宏伟楼梯正上方有着 汤姆士L. 极具代表性的大幅画作；别名寇巴纳的艺术家贾克–孔贝，运用了块体相连的标示性作品来作为各楼层的识别符号；通往客房的各个廊道空间，则采用了红、黄、蓝、绿，彩虹一般的明亮灯光来相互区别。

在楼上，位于中央部分的吧台空间，以60年代相当具代表性的荧光绿为主色，搭配着深灰色的地板、木质深色的桌子，以及以灰红双色调组合的沙发和座椅。酒吧旁边是一个以黑色板岩铺地、带有乌斑木桌椅的餐厅。客房也同样以素洁风格来装修：灰黑色调、超薄荧幕，还有深色原木家具搭配着符合酒店形象的庄重色彩，墙上张铺着印上黑白照片图案的布料，很有50年代的探险风格。

位于圣尼古拉碉堡正下方的柚木制平台，有着面对马赛旧港的绝佳景观。平台旁的游泳池，紧邻着一块从2007年5月起新开辟的葡萄园，种植着酿酒用的西哈品种，也提供了一片绿意。

二重奏酒店
Duo Hotel

Designer:	Jean-Philippe Nuel
Location:	Paris, France
Completion Date:	2006
Photographer:	Agence Jean-Philippe Nuel

设计师：	让菲立浦-努埃勒
地点：	法国，巴黎
完工日期：	2006
摄影师：	Agence Jean-Philippe Nuel

Designing a hotel in the Marais in Paris is not a trivial matter. Like Soho in New York or Omotesando in Tokyo, the neighbourhood is emblematic, and taking the location and the environment into account is fundamental.

The facade is made up of several, typically Parisian, wood-framed shop fronts, remnants of the boutiques that used to be situated here. Inside, exposed beams, several supporting walls that have today been opened up, and the former courtyard of the building transformed into a patio, are various traces of the past on which the present has been built.

The sequence of spaces winding round the interior courtyard allow for a progressive discovery of the place. Each guest can choose the degree of privacy that he or she requires for working, talking or having a drink. As in many hotels, the lobby becomes a living space with multiple functions, abandoning the status of static and representative space. The virtual fireplace, the central library, the bar and the lounges illustrate this new usage. The columns in raw concrete play with varnished wood. Black and white herringbone tweed, a very "couture" statement, stands out and interrupts the aniseed green chromatic range.

The rooms, like the lobby, illustrate a mixture of trends, where patterned wallpapers rub shoulders with contemporary furnishings. The classically inspired furniture is covered in a polyurethane material and mingles with chromed light fittings.

在巴黎的玛黑区设计酒店不是件可以掉以轻心的事，就像在纽约的苏活区或是东京的表参道区一样，玛黑区是巴黎具代表性的区域，将本区的场所环境纳入设计的考量是非常基本而且重要的。

酒店的立面由几个非常典型的巴黎木造店铺门面所组成，令人联想到在此之前存在的商店。在室内，有着外露的木梁结构，有几处原先作为支撑用途、如今被开洞穿通的墙体，还有由以前的天井改建成的室内中庭，当代的设计在这个到处充满历史痕迹的空间环境里被建造了起来。

酒店内部的空间沿着室内中庭而渐次展开，带领人逐渐感受到这里的全貌。人们可以根据其对于私密性的需要来选择工作、讨论或是小酌一杯的空间，就像其他许多酒店一样，该酒店的大厅成为具有复合性功能的生活场所，舍弃了作为特定空间与具代表性角色的身份。其中的装饰性壁炉、中央藏书架、酒吧、沙龙等空间向人们展现了这个新的功能。此外，大厅中的清水混凝土柱与上漆处理过的木料相互搭配衬托，而带着黑白交错图案、非常具有时装风格的织品则与空间中的杏仁绿色调形成饶富趣味的落差。

客房的设计和大厅一样，呈现了多种风格融合一体的感觉，带花纹的壁纸与当代的装修面材并列，古典风格家具以聚酯布料来装饰，并且栖身于金属色泽的灯光照明之下。

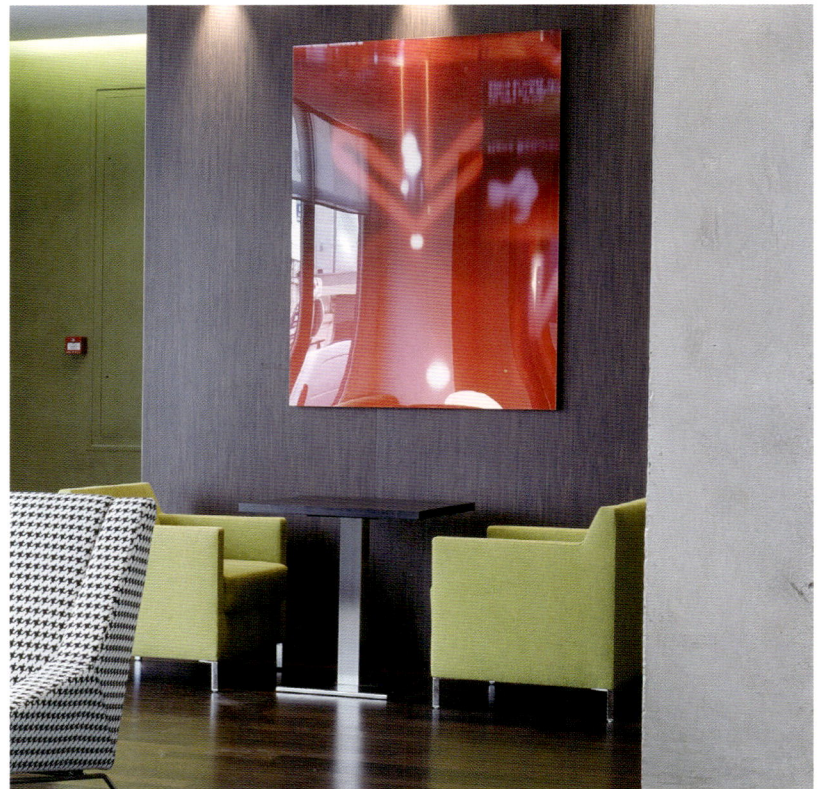

Comfort - Practicality - Dynamism

Contemporary Urban Hotels

都会酒店

舒适—功能—机动

塞尚酒店
Cézanne Hotel

Location:	Cannes, France	地点：	法国，戛纳
Completion Date:	2006	完工日期：	2006
Photographer:	Jérôme Kelagopian	摄影师：	Jérôme Kelagopian

The designer feel of this hotel is epitomised by the bar, upholstered in pale grey leather, with a Perspex counter lit up by the glow of red neon lights, lavastone flooring and splashes of colour on the walls... not to mention a few notes of soul music.

Outside, a raised garden is the setting for a corner terrace in teak with dark furniture in woven resin. Sculptures by Jean-Paul Bongibault blend harmoniously with exotic plants and trees. Mood music fills the air.

The rooms offer a world of colour and warmth. Fuchsia, mauve, blue, red and many other hues brighten every floor. The furniture is in black wood studded with brushed steel. The parquet is in exotic wood. Designer lighting and ornaments add the finishing touches, with leather headboards backlit to show them to their best advantage. The decor in the numerous junior suites features leather sofas and lighting that sets off the glassware...

The gym and steam room have received the same designer treatment, with colours, mirrors and cutting-edge equipment to make sport a pleasurable experience. The steam bath is an invitation to exotic travel with Oriental mosaics and a stone fountain, while beside it is a Zen massage room.

淡灰皮质装修的酒吧、红色霓虹灯照射的玻璃柜台、水洗石材铺设的地板、多彩的墙面，搭配着淡淡的灵魂音乐，这就是充满设计感的塞尚酒店。

在户外，抬高的花园一角设置了一个由木板铺成的阳台，搭配着素雅的深色树脂制家具，在音乐缭绕的氛围中，充满异国风情的植物与让保罗－蒙吉伯的雕塑在此和谐相伴。

在客房方面，出现的是热情多彩的世界，荧光粉红、紫红、湛蓝、赤红等颜色随机分布在不同楼层里，搭配着磨钢沙处理过的乌木家具、异国原木地板，加上精心构思的灯光照明与装饰性陈设，让空间具有强烈的现代精神。从背后打上光线的皮制床头板，显得格外地富有质感；此外，头等套房里的皮质沙发、玻璃饰品和灯光照明，也让空间更显出色。

健身房和土耳其浴室也同样受到精心设计：健身房有着多样的色彩、镜面以及先进的设备，提供了人们舒适方便的运动空间；土耳其浴室则装饰着具有东方风味的马赛克与石造喷泉，仿佛带人进入一场感官之旅，其旁边是充满禅意的按摩室。

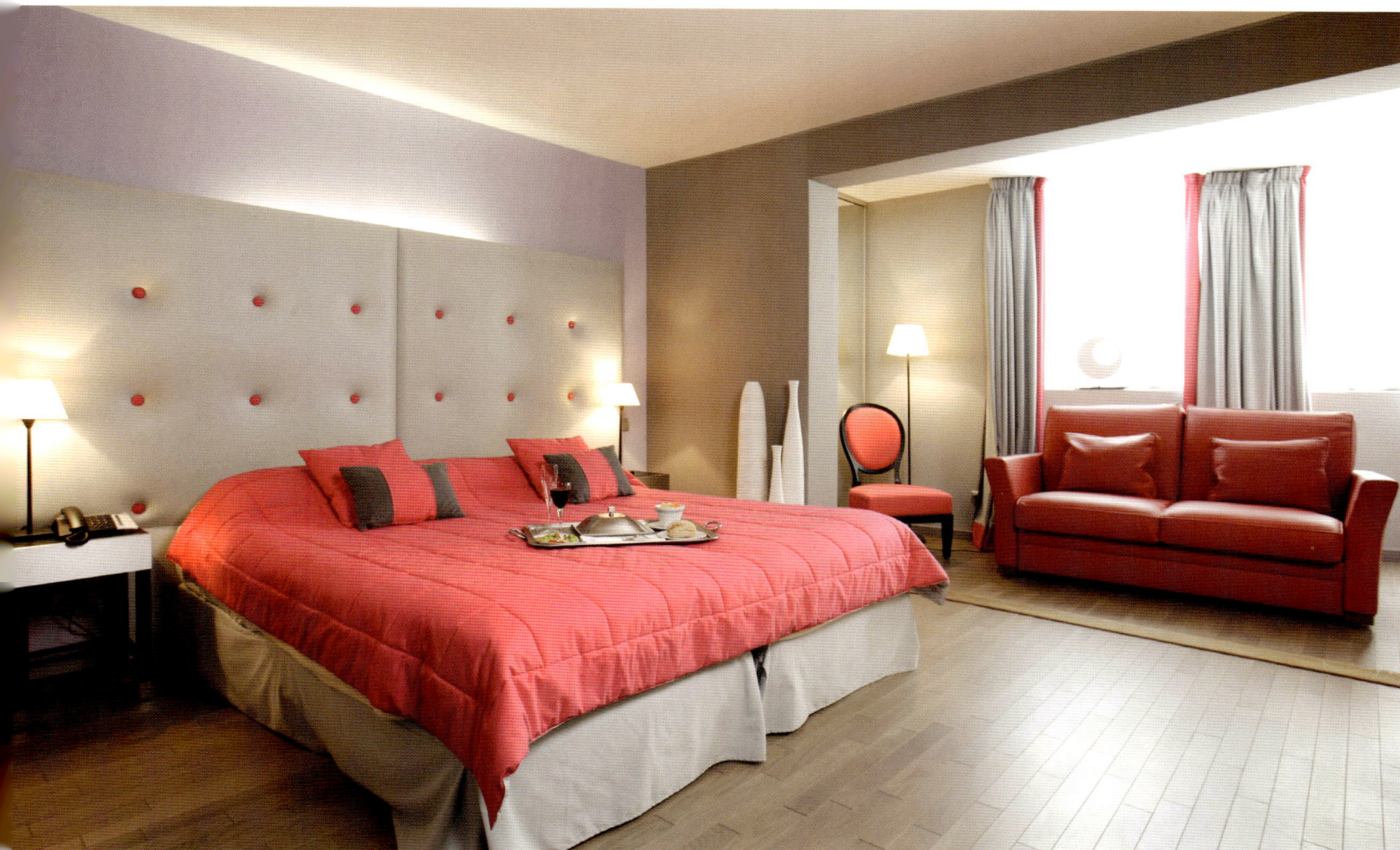

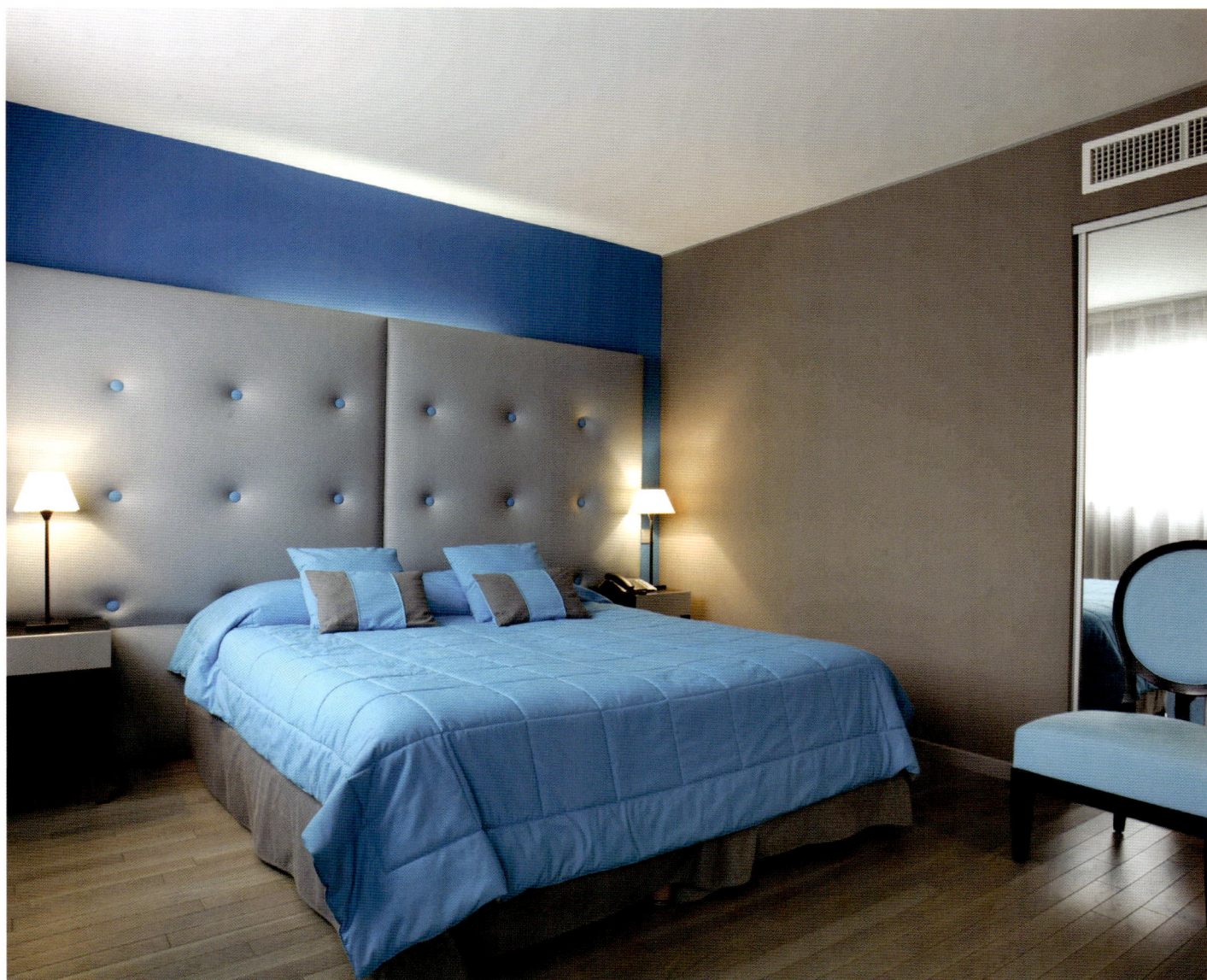

坎博酒店
Cambon Hotel

Designer:	Jean-François Force / Clé Millet Agency
Location:	Paris, France
Completion Date:	2007
Photographer:	Wijane Noree

设计师:	让佛朗索瓦–佛斯 / Clé Millet 事务所
地点:	法国, 巴黎
完工日期:	2007
摄影师:	Wijane Noree

In the heart of Paris, near the Tuileries garden, the Louvre museum and Place de la Concorde, the Cambon Hotel benefits from a privileged situation and attracts an artistically inclined clientele, in an environment that is profoundly contemporary in terms of volumes, materials and specially conceived furniture.

The 40 rooms and junior suites, certain of which are extended by a balcony or a terrace with a view over the Tuileries gardens, have been designed with attention to comfort and the latest technology. This resolutely pure and yet warm environment marries admirably with a remarkable collection of paintings, on view throughout the establishment, as well as several contemporary sculptures. In the reception area, with its sycamore wood sofas, objets d'art, Gallo-Roman pottery, Old Master paintings or even a sculpture by Jean-Louis Raina are mixed in together... Everything has been gathered together over the course of time, according to the designers' desires and passions.

The salon bar, under its glass roof, is decorated with a trompe-l'œil by Marie Joseph Tournon representing the Tuileries gardens, and offers an intimate ambiance in which to relax after a stroll around Paris...

坎博酒店地处巴黎市中心，位于图勒里花园、罗浮宫以及协和广场附近，因为其所在地点绝佳，吸引了不少爱好艺术的旅客。酒店的空间体量、材质和家具都显现十足的当代风格，

在40个一般客房以及头等套房里，有些房间拥有面向图勒里花园的阳台，所有的套房设计都结合了舒适性与最新的科技。在风格简洁亲切的空间里，到处陈列着出色的绘画作品，其中有为数不少的现代作品。在接待大厅里，有着无花果树做成的沙发，交错搭配着艺术品、高卢罗马人的陶艺品、大师画作，甚至有一个让路易–莱纳的雕塑……。这一切装饰都符合了当代精神，呈现欲望和热情。

位于玻璃采光罩下的沙龙酒吧之中有一幅大师玛丽约瑟夫–图尔侬所画的几乎可乱真的图勒里花园风景画作，让人在巴黎漫步之后停留在此的时光更显轻松柔和。

将军酒店
General Hotel

Designer:	Jean-Philippe Nuel
Location:	Paris, France
Completion Date:	2002
Photographer:	Gilles Trillard

设计师:	让菲立浦–努埃勒
地点:	法国, 巴黎
完工日期:	2002
摄影师:	Gilles Trillard

The General Hotel, created by the architect Jean-Philippe Nuel, is situated between the two lively areas of Place de la République and the Marais. Urbane and welcoming, it offers a youthful, exuberant experience with its contemporary and convivial living spaces.

The common parts show a colourful and geometric bias. White plaster columns articulate the volumes of the lobby, while the notion of service is revisited with a varnished rosewood podium integrating all the technological elements of checking in. The decoration of the bar, at the heart of the hotel, plays on contrasts. The white lacquered furniture contrasts strongly with the dark wood of the floor, while the walls have colourful graphic motifs from the 1930s. It is an original place with a relaxed and avant-garde atmosphere.

Each room rests on an individual concept and spatial organisation, dedicated to the comfort and tranquility of guests. The elegant and sober interiors are fitted out to ensure pleasure and rest. The subtle combination of light colours and natural materials such as wood and leather, and the fluidity and modernity of the furniture lines, contribute to this peaceful ambiance.

The General Hotel also offers its guests a functional and flexible space for work and meetings. Finally, a fitness centre offers the chance to relax in the gym or sauna.

由建筑师让菲立浦–努埃勒所设计的将军酒店位于巴黎两个极为热闹的区域之间：玛黑历史街区以及巴士底广场街区。同时具有城市特性与热情风格的将军酒店，让人们在此经历一种对生活空间的发现之旅，是一个具有现代感与社交感的场所。

酒店公共空间的设计体现出一种强调色彩与几何形体的建筑主张：白色的柱列将酒店大厅的几个不同空间体量整合起来；上漆的红木柜内部暗藏着录音设备，完美地结合了具现代感的优雅装饰与物件的功能性。服务的概念在此被重新思索，透过设计而展现。位于酒店中心的吧台空间以强化对比性为设计主调，经过漆艺处理的白色家具被深色的原木地板衬托得十分鲜明出色，墙上则铺陈着1930年代风格的图案。这个场所释放出一种轻松却前卫感十足的空间氛围，因此显得独具特色。

酒店里的每一个房间都具有独特的设计概念以及空间安排，并且着重于为旅客们提供舒适与宁静。因此，室内的装修以优雅简洁为主调，让人们的视觉拥有美观愉悦之感，也同时可获得休息。清浅色调与自然材质(如原木与皮革)的巧妙结合，家具线条的流畅性与现代感，皆为室内空间带来平和宜人的氛围。

将军酒店还提供了一个功能性极佳的工作与会议空间，能够弹性适应客人的需求。此外，它也设有能够让人消除疲劳、运动健身的养生中心和桑拿浴空间，使旅客在优美细致的环境中充分休息之后再重新出发。

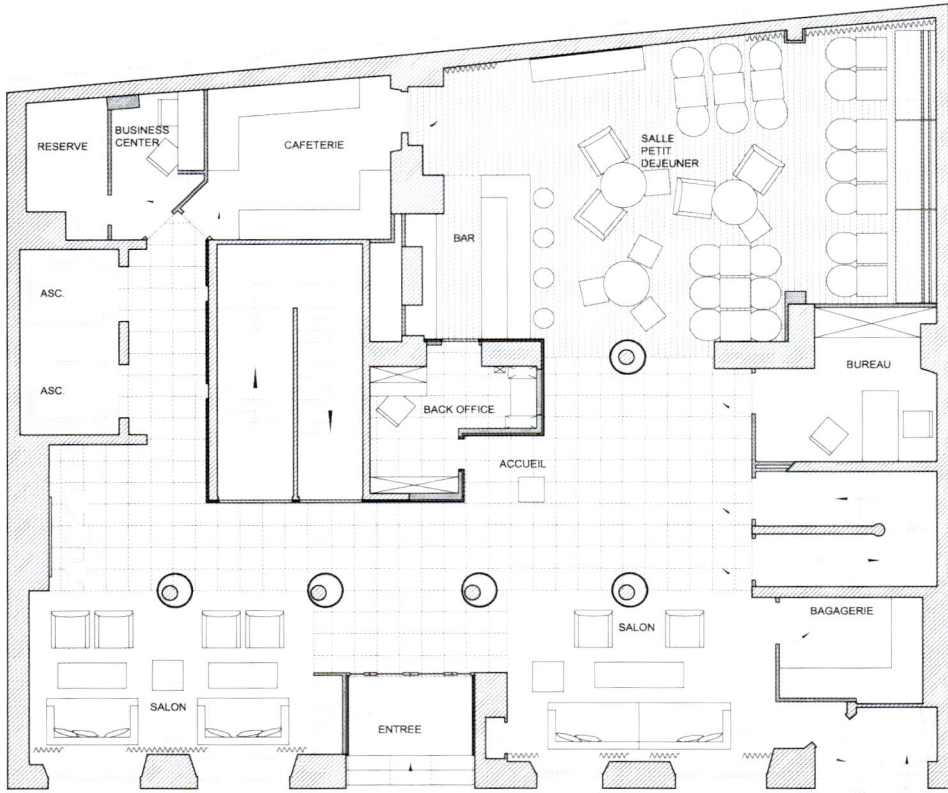

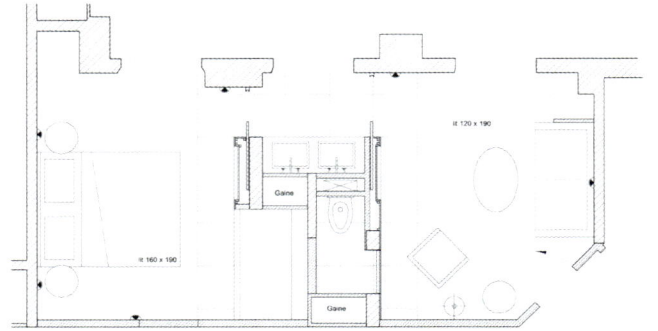

葛兰德酒店
Le Grand Hôtel

Designer:	Jacques Molho
Location:	Strasbourg, France
Completion Date:	2007
Photographer:	ACTWEB

设计师：	贾克-莫侯
地点：	法国, 斯特拉斯堡
完工日期：	2007
摄影师：	ACTWEB

The Grand Hôtel's history goes back to the beginning of the 18th century. Entirely destroyed by the bombardments of the Second World War, it reopened its doors in 1954. Its panoramic lift offers a view of the magnificent 6-floor staircase, both of them dating from after the War. Revolutionary in its day, the lift was brought from the United States by Dreiber, an Alsatian architect. Still looking utterly contemporary, it brings an additional charm to a spacious and luminous whole. Important renovation works started in 2006, giving birth to a design hotel. Since then, the hotel has enjoyed a new reception area, entirely conceived in modernist and original tones.

Modern and contemporary, the Grand Hôtel plays on its materials–wood, natural stone–and lighting, thus creating an atmosphere that is both Zen and welcoming. The reception area and its salon-bar with its own fireplace invite in and seduce with the clarity of its pure lines. Original works by Ayline Olukman, a young Strasbourgeoise artist, are displayed here.

The 83 rooms combine pure lines and warm colours such as brown, ochre or plum, for a carefully thought-out decoration. The bedrooms are all different but nevertheless in the same spirit. The refinement of orchids, the choice of lighting, of an accessory, bring a subtle taste to the bespoke decoration. The bathrooms, all in marble, offer their own tints with nuances of white, red, ochre and sky blue...

创造于18世纪初的葛兰德酒店，在第二次世界大战期间被炮火完全摧毁，之后又在1954年重新开幕。酒店的观景式电梯让人能够在此纵览一座高6层楼的壮观大楼梯，电梯与楼梯两者都建造于战后。这座电梯在建造当时是十分前卫的产品，由法国西边阿尔萨斯地区的建筑师德瑞巴从美国带回来的，这座电梯如今依然不失现代感，让这整个明亮宽敞的空间更增添了独特迷人的风采。葛兰德酒店在2006年开始进行一系列重要整修工程，使其更成为一个充满设计感的酒店，特别拥有了一个具有现代主义色彩的新颖接待空间。

崭新现代的葛兰德酒店刻意采用不同材料来进行装修：原木、自然石材，加上光线的巧妙运用，使得这里的空间带有一种亲切、饶富禅意的特质。接待空间和旁边带有壁炉的沙龙酒吧以其明亮特质与简洁线条吸引着过往的旅客，这里还展示着年轻的阿尔萨斯艺术家艾琳-欧露克曼的作品。

结合了简洁线条与棕、赭石、紫红等暖色系列色彩的83个房间展现出一种考究的装饰风格，这些设计各不相同的客房却同时保有一致和谐的气质。兰花的细致高雅、灯具的选择、配件的讲究，都确保了整体空间的别致品位，像是特别为这里量身打造的一样。有一些浴室甚至使用了白色、红色、赭石色、天蓝色……等花色不同的大理石来设计。

The breakfast room assures the continuity of the design theme with discreet and refined tones of plum, chocolate and sand. When the good weather comes, the shady teak terrace offers the possibility of having breakfast in a lazy and escapist atmosphere... Tables that are alternately in dark wood for the interior and white for the terrace, plum coloured chairs and small oval lamps filtering the light, reinforce the tranquility of the setting.

早餐厅使用了紫红、咖啡和栗子色来延续酒店的整体设计感以及其雅致庄重的风格。在风和日暖的时节，铺着木质地板的阳台提供旅客一个慵懒、远离尘嚣的早餐时光。室内的深色原木餐桌和阳台上的白色餐桌，皆搭配着紫红色的座椅，加上椭圆灯罩散发出柔和光线，空间显得更为祥和宁静。

帝国酒店
L'Empire Hotel

Designer:	Roland De Leu	设计师：	罗兰–德勒
Location:	Paris, France	地点：	法国，巴黎
Completion Date:	2007	完工日期：	2007
Photographer:	Wijane Noree	摄影师：	Wijane Noree

L'Empire Hotel combines the 18th-century style of its facade, an architectural symbol of this epoch, and the pared-down modernity of the 21st century. Here vegetation rubs shoulders harmoniously with the frenzy of the urban environment. The Leitmotiv chosen by the owner and the architect was to mix French sophistication with an atmosphere of contemporary architecture. Behind the large blue-lacquered carriage entrance hides a striking hall where past and present meet and suggest surprises to come.

Over six floors, unembellished grey corridors lead to exceptional rooms and suites. The astonishing volumes immediately give a feeling of ease, space and serenity. The decoration of the rooms, which enjoy remarkably high ceilings, subtly links the whiteness of the walls, bed covers and cupboards with the dark colour of the floor for a very contemporary result. The refined and unusual furniture, such as the giant bedheads in solid wood and beige quilted leather, create a special harmony with the wenge wood parquet present in all the rooms. The "Joséphine de Beauharnais" suite and the "Bernadotte" deluxe room have a private terrace bordered with bamboo, which gives them a great luminosity as well as an opening onto the exterior.

L'Empire Hotel is built around three interior gardens which offer up their spaces in different tones: Blue, Yellow and White, in subtlety and light.

巴黎帝国酒店的建筑立面结合了正统典型的18世纪风格以及21世纪的纯粹现代风格，绿化空间和都市繁忙的环境两者在此和谐的交融。业主和建筑师的设计动机在于将法式的生活艺术和当代的建筑环境紧密地结合起来，在经过漆艺处理的蓝色大拱门之后，抵达了迷人的接待大厅，在这里传统与现代两种风格融合交会，暗示人们接下来将会有的空间惊喜。

客房分布在六个楼层，旅客们经由银灰色、造型简洁的典型廊道而前往各个房间。房间内令人惊异的宽广量让旅客们立即感到自在而放松。大于一般酒店房间的空间高度、雪白细致的墙面、精美的床罩以及衣橱，都与深色的地板相互衬托、经营出一种时髦的现代感。细致而独特的家具一例如用厚实木材以及羊毛灰色真皮制成的巨大床头板一和所有房间的原木地板相互衬托，让整体空间呈现出一种特别的和谐效果。名为"约瑟芬–德–博阿尔内"的高级套房以及名为"贝尔纳多特"豪华客房各自拥有一个种植着竹丛的私人阳台，不仅使室内空间更加明亮，同时可享有对外开放的视野。

巴黎帝国酒店的内部空间围绕着三个不同色彩主题的庭园而分布：湛蓝、鹅黄和雪白，这三种花色让旅馆的空间显得更为细致轻盈。

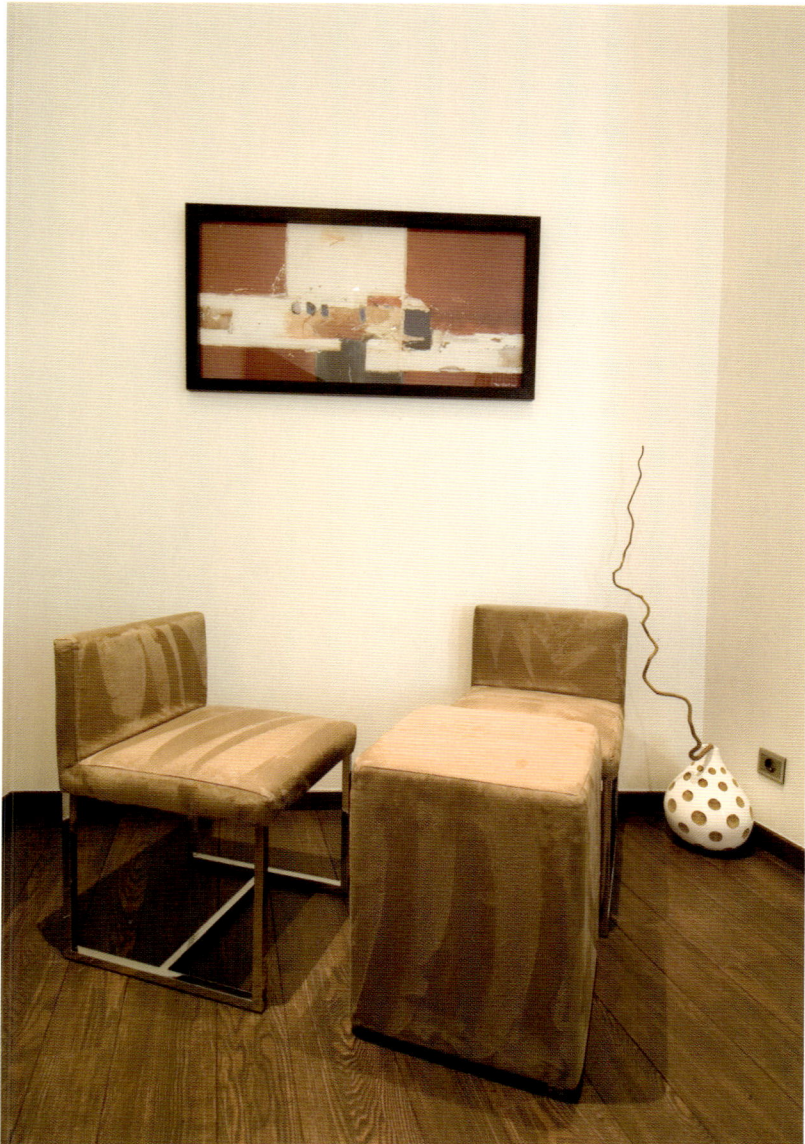

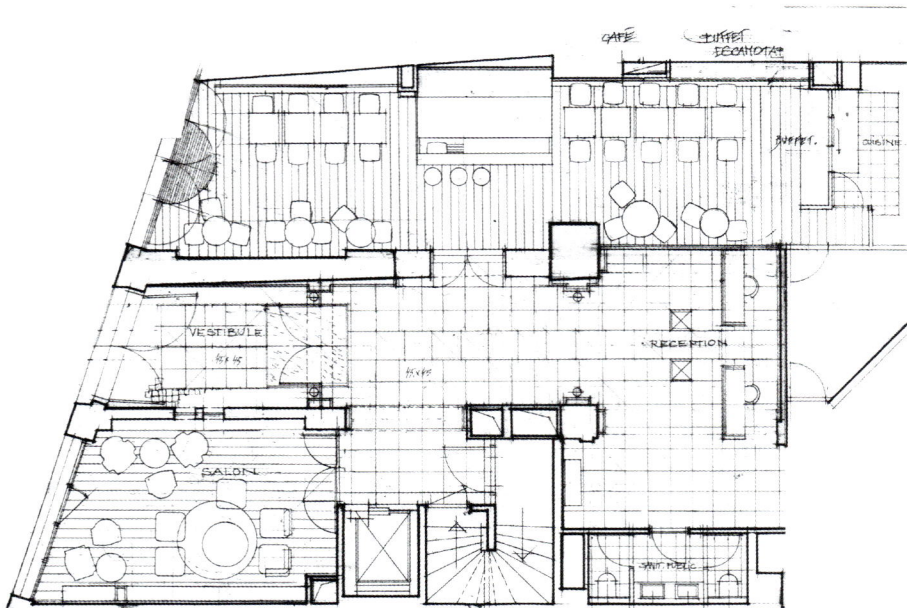

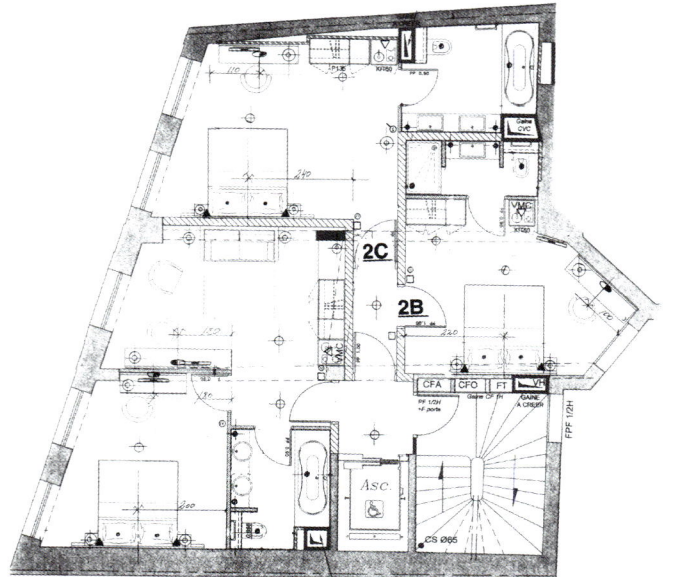

子午线酒店-巴黎星辰广场店

Le Méridien - Paris Étoile

Designer:	Pierre-Yves Rochon
Location:	Paris, France
Completion Date:	2001
Photographer:	Le Méridien - Paris Étoile (pp.120, 122, 123),
	Wijane Noree (pp.121, 124-127)

设计师：	皮埃尔伊伊夫–罗逊
地点：	法国, 巴黎
完工日期：	2001
摄影师：	Le Méridien - Paris Étoile (pp.120, 122, 123),
	Wijane Noree (pp.121, 124-127)

The Hotel Le Méridien Étoile, built in 1972, was entirely renovated in 2001 by Pierre-Yves Rochon. A mixture of French tradition and refined modernity, the Méridien Étoile sees itself above all as a cosy and luxurious place to stay, open to the world and its influences.

The lobby gives a foretaste of what is to come. It opens onto the different spaces of the hotel to allow a natural fluidity, a veritable game of transparence and space. This impression is reinforced by the choice of colours and by symmetry. Decorative elements such as Chinese vases, Eastern-influenced rectilinear lamps, furniture with modern lines, paintings by contemporary artists… cultivate the atmosphere of a resolutely chic and modern interior.

The same refinement is found in the 1025 rooms and 21 suites. The silvered room numbers are inspired by Art Deco style. The gentle tones of beige, nuanced on the walls and above the bed, and flecked beige for the carpet, are combined with the essences of warm wood. Chests of drawers in dark wood or acacia, depending on the floor, sycamore bedheads, black lacquered furniture, armchairs with a footrest and round tables for the lounge area, harmoniously complete these cosy rooms. The bathrooms sport modern lines: black granite for the floor, white glazed tiles for the walls and a washbasin in black marble. Lithographs, reproductions and photographic montages decorate the walls, a nod to jazz and to "Right Bank Paris".

建造于1972年的巴黎星辰广场(凯旋门)子午线酒店，在2001年由建筑设计师皮埃尔伊夫–罗逊全面重新整修。融合了法兰西传统与细致现代感的子午线酒店，意图成为一个豪华舒适生活的代表性场所，一个向全世界开放的场所。

呈现酒店第一印象的大厅，同时对着各种不同空间开放，形成自然流畅的空间关系。这种空间的通透印象借由色彩的搭配以及对称设计手法而更受到强化。大厅中的各种装饰，如中国式的瓶樽、东方的四角灯笼、现代线条的家具、当代艺术家的作品……，都让空间显得更高贵典雅而不失现代感。

酒店1025个标准客房和25个豪华套房的设计也呈现出同样的细致品位，房间号码的银色字形设计灵感来自于20世纪初装饰艺术的风格，壁面、寝具、地毯都是以乳灰色为主调，以不同深浅的层次变化来搭配家具温暖的原木色系。依楼层不同而深浅有异的原木橱柜、无花果木制成的床头板、黑色漆艺处理过的家具、带着取暖矮椅的扶手沙发以及沙龙角落的小圆茶几，一起和谐地构成了客房的柔和舒适感。以现代概念设计的浴室有着黑色花岗岩的地板、纯白色的壁砖以及黑色大理石的洗手台。墙上挂着几幅让人联想到爵士音乐或是巴黎河右岸气氛的石印图画与摄影作品。

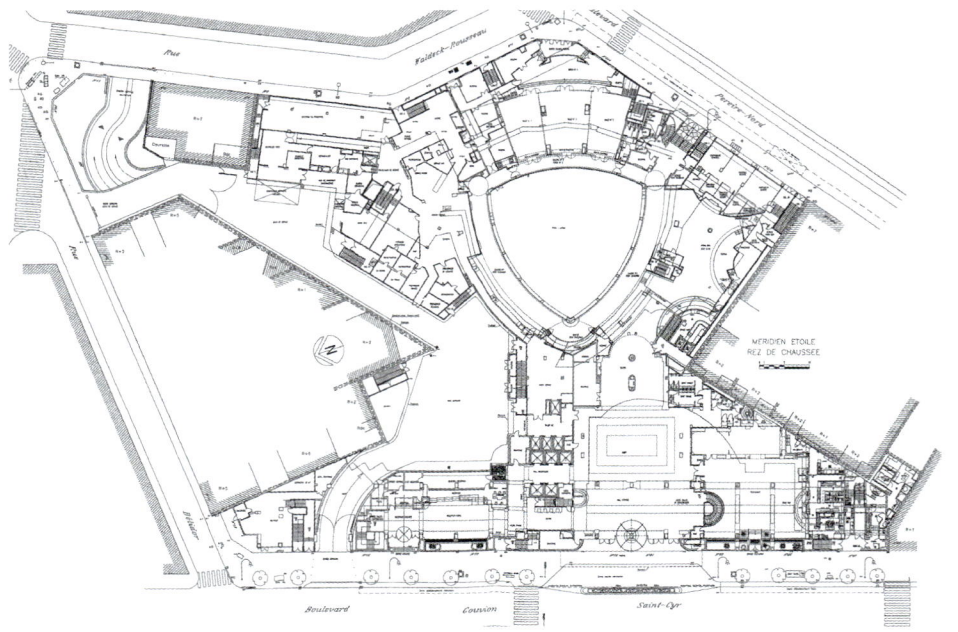

As for the restaurants, the Terrasse du Jazz adopts the tones of aniseed green and celadon. The large picture windows of the restaurant give on to the hotel's interior garden where a small waterfall brings sweetness and serenity. At L'Orénoc restaurant one finds a welcoming and timeless atmosphere, where the invitation to travel is suggested in the smallest decorative detail: tones of terracotta and essences of rare woods serve as a background to conversation about past voyages and those to come.

在餐饮空间方面，名为爵士露台的餐厅以茴香淡绿为主色，大片玻璃窗面对着酒店内庭花园中的阶梯式喷泉，让人倍感宁静温馨；罗黑诺餐厅带给人们一种充满热情、超越时间的感觉，其空间装饰的每个细部都再次邀请人们踏上旅程：不同陶土的色调与各种稀有木材的质地，皆成为反映着过去或未来之旅的藏宝盒。

凯伯乐酒店
Keppler Hotel

Designer:	Pierre-Yves Rochon	设计师：	皮埃尔伊夫－罗逊	
Location:	Paris, France	地点：	法国，巴黎	
Completion Date:	2007	完工日期：	2007	
Photographer:	Fabrice Rambert	摄影师：	Fabrice Rambert	

The Keppler Hotel offers 34 rooms and 5 suites, of which some have a terrace and a superb view over the rooftops of Paris and the Eiffel Tower. The windows have black blinds bearing a signature white K. The lower part of the facade is in lacquered wood, allowing a glimpse of one of the hotel's lounges. The interior architecture by Pierre-Yves Rochon is a subtle mix of classicism and contemporary creativity. The diversity of styles and the richness of materials create a delightful harmony.

The reception is adorned with a large carpet with asymmetrical black and white motifs, on which a large leather-covered desk presides. The two lounges with a parquet in American walnut and beautiful moulded ceilings, one with a glass roof, exude a warm atmosphere. The Library Bar, dotted with Rothko, Picasso and 1930s engravings gives onto a garden with an intimate atmosphere.

Breakfast is a privileged moment in this winter garden, which brings it a touch of magic. White medallion chairs and black lacquer tables give character to the decoration of this green-walled room where the delicious breakfast buffet is laid out. The open kitchen, decorated in black lacquer and zinc, adds a pleasant conviviality.

凯伯乐酒店一共有34间标准客房以及5间豪华套房，其中的一些房间拥有居高临下俯瞰巴黎屋顶或是面对埃菲尔铁塔的观景阳台。旅馆的窗户采用了黑色窗框搭配白纱窗帘，立面的下半部是由经过黑色漆艺处理的木板装拼而成，来往的行人可以透过立面的开口看见酒店内部的一个沙龙。由皮埃尔伊夫－罗逊设计的室内空间细致地融合了古典精神与现代创意，在风格的多样化以及装修材料的丰富性之下，达到整体的和谐效果。

酒店接待处以黑白双色不对称的地毯来铺地，其上设置着一张皮质的大书桌。旅馆的两个交谊沙龙用的是美洲胡桃木的地板，搭配着饰有精致雕花线脚的天花板，其中一个沙龙覆盖着采光玻璃罩，进到室内的温柔光线让这里充满亲切迷人的气氛。图书酒吧空间的墙上挂着几幅罗斯科、毕卡索以及30年代的版画，衬着旁边的花园，让这里充满温馨柔和的氛围。

在这个温室花园里有着白色圆形靠背椅和黑色漆质餐桌，沿着绿色的墙面延伸，成为空间的特色，早晨在这里享用自助式的早餐将会是令人难忘的时刻。一个由黑色漆质和镀锌材质装修的厨房对着这个空间开放，为此花园餐厅带来一种真实的生活感。

The rooms are accessed by wide corridors. The doors in dark wood are framed in sculpted medallions. Some suites enjoy a large teak terrace with a view of the Eiffel Tower and the Arc de Triomphe. The marriage of revisited Louis XV and Louis XVI style gives these rooms elegance, comfort and refinement. The walls and curtains have a base of black and white embellished with touches of colour: brown and red for the suites and yellow, mauve or sea green for the rooms. An English-style refinement unfolds in the bathrooms with walls dressed in toile de Jouy, waxed or back-striped fabric. Black porcelain taps, floors in black marble and plasma screens make these rooms the epitome of comfort.

And, under a vaulted ceiling with exposed stone, a fitness room and sauna offer a delicious moment of relaxation and well-being.

宽广的廊道带领旅客前往客房，原木材质制成的房门镶嵌在雕饰精美的门框中。若干豪华套房具有原木铺置的私人阳台，让在此住宿的旅客可以欣赏面对埃菲尔铁塔或是凯旋门的景观。房间内摆设着以路易十五和路易十六风格为基础重新设计的家具，让空间显得更优雅、舒适而细致。所有的客房的墙面都以黑白双色为基调，在豪华套房里加以红色或是深紫色来点缀搭配，一般客房里则使用鹅黄、淡紫或是水绿。浴室里的装修采用了英格兰的细致风格，蜡纸或是黑白条纹式壁纸的墙面衬托着黑瓷的水龙头、大理石的地板以及液晶荧幕，使得此空间仿佛一个舒适的宝盒。

此外，位于拱形石造天花下的健身房与桑拿洗浴中心，能够让人在此放松休息后，重新恢复身心最佳状况。

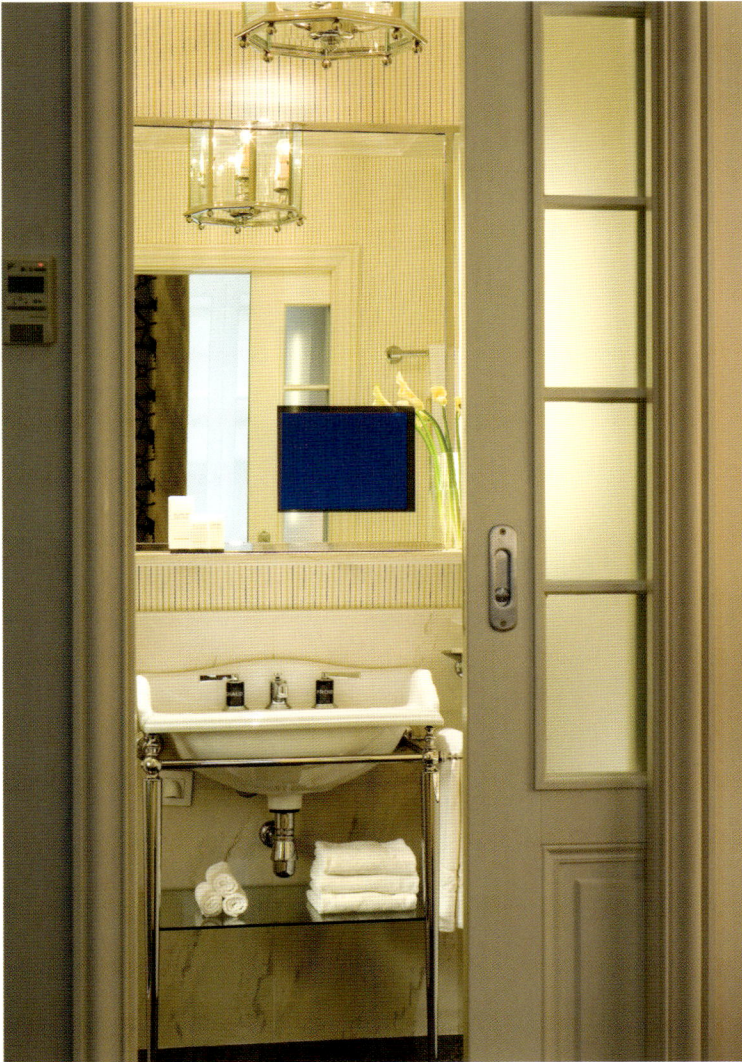

乐维农酒店
Le Vignon Hotel

Designer:	Jacques Mechali
Location:	Paris, France
Completion Date:	2006
Photographer:	Fabrice Rambert

设计师：	贾克－梅查理
地点：	法国，巴黎
完工日期：	2006
摄影师：	Fabrice Rambert

Peaceful and welcoming, the hotel Le Vignon entices one away from the Parisian tumult. Situated in a road where it is pleasant to walk, Le Vignon, with its 28 rooms, mixes Parisian cachet and contemporary style, in the heart of the "fashion quarter" of Madeleine.

Comfortable sofas, joyful colours, and discreet and serene fragrances come together here to create a welcoming atmosphere. A lounge on the ground floor is perfect for business meetings as well as to have a drink at the end of the evening.

The rooms are warm and elegant. The beds invite one to rest, and the bathrooms to relax. The double curtains in white canvas printed with yellow and raspberry give the Superior rooms a happy and restful atmosphere, and a cocooning feeling of comfort marries with subtle nuances of grey in the De Luxe rooms on the top floor, which are chic and pure of line.

Designed as a Parisian apartment in which one can occupy a room or the whole space, the Top Floor has a resolutely contemporary feel, prioritising silence and light. Under the eaves, it offers a poetic space. Like a precious box, the bedroom is ideal for both work and relaxation with its corner for afternoon tea...

地处喧嚣的巴黎市区中，乐维农酒店平静与热情的性格显得更加诱人。这家拥有28个房间的酒店位于玛德莲教堂时尚街区的中心，一条适合散步闲逛的街上 (即乐维农街)，融合了巴黎的传统特色和当代的风格。

舒适的沙发、悦人的色彩，加上令人心神平静的清淡香味，塑造出这里亲切宜人的氛围。地面层的沙龙提供人们进行商务会谈的空间，同时也是让旅客可以在晚上小酌一杯的场所。

所有客房都具有优雅而温暖的性格，有邀请人静心休息的床铺、让人身心放松的浴室。高级客房在白色帷幔为底、衬上红黄双色布帘的双层窗帘的衬托下，显现愉悦与平和的气氛。位于顶楼的豪华客房则采用不同层次变化的灰色系列，来搭配柔和舒适的空间，表现出高雅纯净的风格。

顶楼豪华客房被设计成像是巴黎的私人公寓一般，有着专为个人设计的房间或是整体使用的大空间，并特别注重光线与宁静，十分符合时下趋势。这个屋顶下方的空间，是极富诗意的场所，此处的客房有如珍贵珠宝盒一般，适合旅客们在此工作，并在专属的"茶与咖啡空间"中放松休息。

柏悦酒店-巴黎凡登广场店
Park Hyatt – Paris Vendôme

Designer:	Ed Tuttle	设计师：	埃迪-塔投	
Location:	Paris, France	地点：	法国，巴黎	
Completion Date:	2002	完工日期：	2002	
Photographer:	Fabrice Rambert	摄影师：	Fabrice Rambert	

The interiors by Ed Tuttle put the Park Hyatt Paris-Vendôme in a league of its own. His vision was to build a hotel that reflected the pristine qualities of Parisian lifestyle, using fine craftsmanship to create a timeless aura throughout.

The entrance on rue de la Paix opens onto a striking limestone foyer that sets the stage for Tuttle's simple, modern use of classic French materials. Artwork has been carefully selected for display throughout the hotel, making it a showcase of sorts for contemporary Franco-American art.

The restaurant-lounge "Les Orchidées" lies at the heart of the hotel and features dramatic colonnades, gilded ceilings and custom designed silk-cotton chenille. Two courtyards – one glassed over, the other, called "La Terrasse", in the open air – are separated by a large fireplace. The tables of the second restaurant, "Le Pur'Grill", are placed around a raised silver-leafed dome in the centre of the room, which is surrounded by a colonnaded rotunda.

Continuing the theme of understated simplicity, guestrooms and suites have been decorated in neutral tones with exquisite artistic elements. The two-story Presidential Suite's imposing entrance, six meters high and wide, on the sixth floor opens onto an ultra luxurious haven featuring two fireplaces, stone staircases, parquet floors and three small terraces.

位于巴黎凡登广场的柏悦酒店，其室内空间由建筑师埃迪-塔投所设计。埃迪-塔投企图以此酒店的设计来反映巴黎生活风格的典型特色，运用精致的工艺技巧塑造一种永恒的气息。

位于和平路上的酒店入口，将人带入一个引人注目的石灰岩门廊，此门廊为设计师运用法国古典材料的简洁、现代手法提供了舞台。在酒店里的装饰艺术品都经过精心的挑选，让人觉得这里仿佛是美、法当代艺术的展示橱窗。

位于酒店中心、名为"兰园"的餐厅酒吧有着戏剧性的柱廊以及镀金天花板，并搭配着丝棉混纺的设计织品。其中的两个中庭由一座巨大的壁炉区隔开来：一个在玻璃采光罩下的半户外中庭，另一个是被称作"大平台"的露天中庭。酒店的第二个餐厅"普瑞烧烤餐厅"围绕着位于空间中央、饰着银叶的高耸圆顶而设置，此天顶的周围环绕着圆形柱廊。

这个简洁的空间风格在客房的设计上继续延伸，设计师在此运用了中性色调以及精美的艺术元素来进行装修。位于酒店7楼，占地两层楼的总统套房有着宽高各6米的气派入口，引人进入华丽的天堂，里面有着两座壁炉、石造楼梯、原木地板，还连接着户外的三个小巧的阳台。

Bathrooms are an important feature in all the rooms and feature sliding mahogany floor-to-ceiling "disappearing doors". For additional pampering, Le Spa is available exclusively for hotel guests. There are also six rooms available for private functions and meetings, each of which has been designed to correspond with its namesake precious or semi-precious stone.

在客房中占有极重要地位的浴室，从天花到地板都采用了胡桃木来装修。这里的水疗养生中心(Spa)仅仅提供住在酒店的宾客使用，而不对外开放。此外，酒店还有6个专为会议或是其他私人用途所设计的空间，每一间都以一种宝石或半宝石来命名，其内部空间便是以同名的石材来装修。

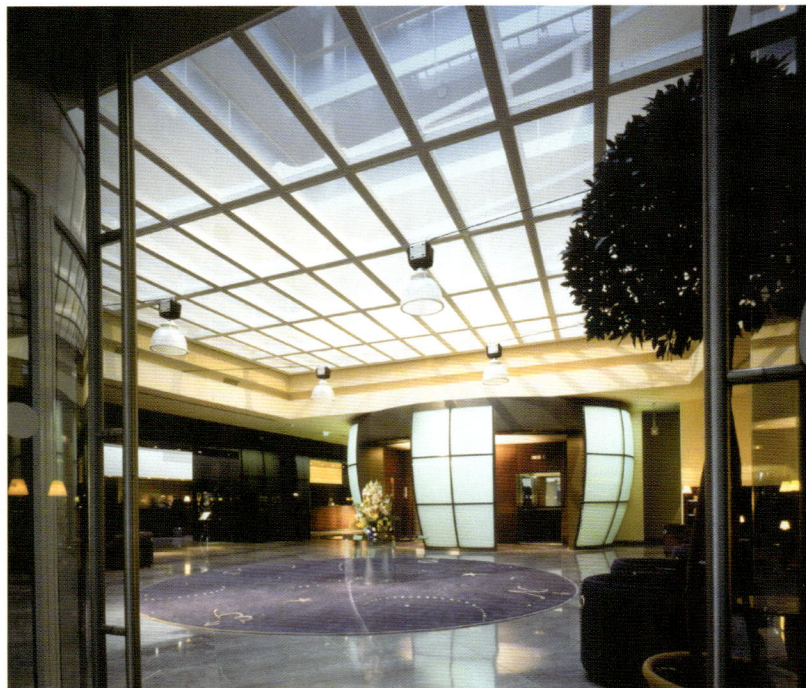

喜来登酒店 – 巴黎机场店
Sheraton - Paris Airport

Architect:	Paul Andreu
Designer:	Andrée Putman
Location:	Paris, France
Completion Date:	1996
Photographer:	Fabrice Rambert (p.148), Wijane Noree (pp.149-153)

建筑师:	保罗–安德鲁
设计师:	安德蕾–普特曼
地点:	法国, 巴黎
完工日期:	1996
摄影师:	Fabrice Rambert (p.148), Wijane Noree (pp.149-153)

Inaugurated in 1996, the hotel rises like an ocean liner at the heart of Roissy Charles de Gaulle airport's Terminal 2. The hotel was built by the Compagnie Générale du Bâtiment et de Construction with Paul Andreu and the architects of the Aéroports de Paris as project managers.

A system was specially developed to filter the noise of the planes. The exterior windows are equipped with triple glazing. Between the windows is an air pocket filled with a sound-absorbing casing which soundproofs up to 45db (A). Aside from this, the hotel rests on a platform made up of two concrete floors separated by spring boxes. This ahead-of-its-time system allows one to filter all the noise pollution.

In addition, a 4 cm Carrara marble cladding and an 80 mm covering of rock wool insulate the concrete parts of the hotel. Thanks to these innovative techniques the Sheraton Paris Airport hotel was voted the "quietest airport hotel in France" in 1998 by the Paris Tourist and Convention Bureau.

The interior decoration as well as the choice of amenities was entrusted to Andrée Putman. This grande dame of decoration succeeded in creating an atmosphere straight from the world of Jules Verne while providing a welcoming and cosy ambiance evocative of travel in each of the hotel's 254 rooms and suites. Also worthy of note are the luminous, brightly shining stars in all the hotel corridors which bring a relaxing atmosphere to its atrium.

在1996年启用的喜来登酒店以一艘船般的姿态矗立在巴黎华西戴高乐机场第二航站特区的中心，由建筑师保罗–安德鲁以及巴黎机场建筑规划中心的设计师共同设计，工程则由房屋建筑与营造联合公司来负责。

因为地处机场的缘故，设计师特别研究发展出一套能够解决飞机起降噪音问题的技术装置：外层窗户具有三重玻璃，而在内外窗户之间则配备有一个能够将噪音减低45分贝的真空层。构造上的另一个特色是：整栋建筑物是建造在由两层分离的混凝土板组成的基础上，并且借用防震箱将它们互相隔离，这个先进的系统使得建筑物能够更佳地隔离外部噪音。

此外，在建筑物的混凝土外墙上采用了4厘米厚的大理石壁砖，加上其后还铺有一层8厘米的岩棉来作为隔离材料。巴黎华西机场喜来登酒店在1998年被巴黎旅客中心与巴黎议会中心选为法国最宁静的机场酒店，便是归功于这个高科技的运用。

室内设计以及家具设备的选择则由室内设计界的女性大师安德蕾–普特曼负责，在旅馆的254个房间与套房里，她以儒勒–凡尔纳作品中所塑造的冒险旅游世界为灵感泉源，成功地在这里塑造出热情好客的空间气氛。酒店各处的走道空间有着像是夜空繁星熠熠闪烁的天花板，格外为中庭空间带来休闲轻松的氛围。

巴士底街区酒店
Le Quartier Bastille Hotel

Designer:	Pascale Douillard
Location:	Paris, France
Completion Date:	2006
Photographer:	Gilles Trillard

设计师：	帕斯卡–杜亚尔
地点：	法国，巴黎
完工日期：	2006
摄影师：	Gilles Trillard

In her design concept for the Le Quartier Bastille Hotel, the architect Pascale Douillard played on the contrast between the positioning of the Le Quartier hotels, which is very contemporary, urbane and Parisian, and the presence of an exterior space which had previously been left virtually unexploited. One of her prime objectives was to optimise the well being not only of clients, but also of the staff, the pleasure of one having an effect on the other and vice versa. She therefore redesigned the volume of the ground floor to give a feeling of fluidity, clarity, openness of the space, and integrating this prolongation towards the outside, which was transformed into an island of nature, Japanese in style, thanks to a fountain and jasmin trees, etc.

Besides the conception of the spaces, this dialogue between the two extremes of city and nature unfolds in the choice of materials, colours and furniture, in particular that designed exclusively for the hotel, such as the bedroom furniture and the breakfast tables... This dialogue is clearly seen in the choice of digitally printed photos for the reception and restaurant representing the Bastille column and vegetation, and is found again in a more subtle form as a theme running through the fitting-out and the decoration. Thus the lines that one finds in the hotel are strict and pure, while at the same time being gentle and curved, in a harmony of subdued hues enlivened by several notes of vibrant colour. City and nature !

在巴士底街区酒店的设计当中，建筑师帕斯卡–杜亚尔企图将非常具有现代感、巴黎都会感的酒店本身与其周围的户外空间形成一种对比。她的主要设计目的之一是让旅客在此获得最大的身心舒适，同时也兼顾酒店服务人员的需求，她认为两者的愉悦是相辅相成的。她将地面层的空间重新组织，在与户外空间相互融合之下，塑造出更大的流畅性、明亮感和开放性。在户外空间的处理上，因为喷泉、茉莉花等元素的运用，让人有如身处日本自然庭园的感受。

这种界于自然和城市两极之间的对话，除了在空间的设计概念中展现出来之外，也体现在材料与色彩的选择中，尤其是为酒店特制的家具，如客房里的桌子、早餐桌……等。这个对话也同样出现在接待处与餐厅墙面上数位打印照片影像的选择，分别呈现的是巴士底广场圆柱和植物的影像。这个对话概念在其他空间的设计与装饰之中有如主导原则，被更细致婉约地诠释出来。人们在酒店里看到的是孕育在柔和氛围与弧形线条之中、严谨而纯净的风格；在和谐温婉的整体色彩之中偶尔出现鲜丽的颜色来点缀，再次呈现都会与自然对话的主题。

PHOTO Denis FELIX

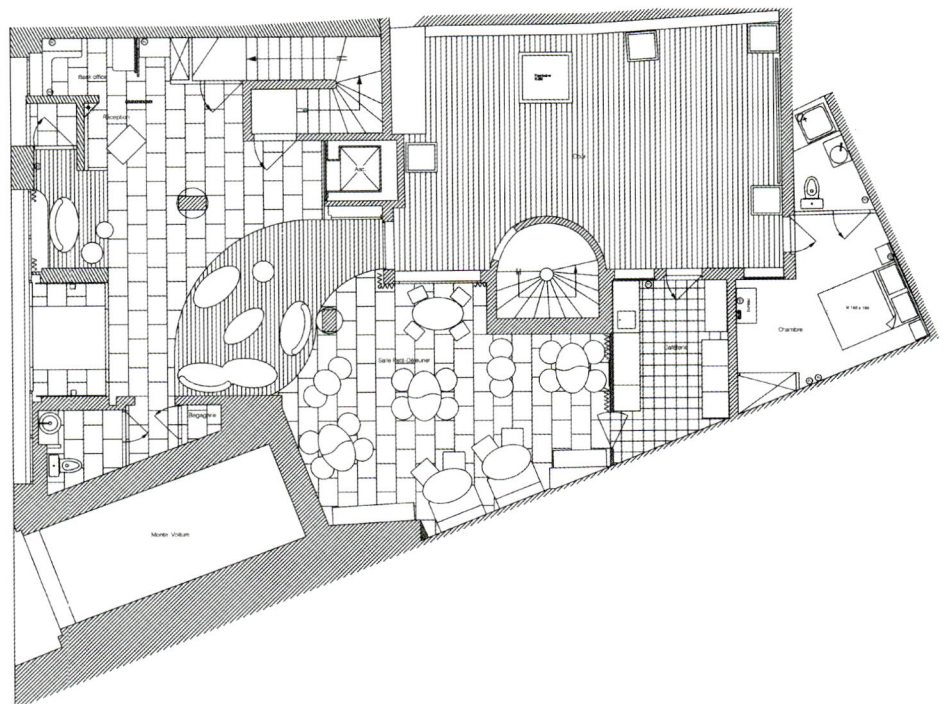

影像工作室 酒店
Les Ateliers de l'Image

Designer:	Roland Paillat	设计师：	罗兰-裴拉
Location:	Saint-Rémy-de-Provence, France	地点：	法国, 圣雷米-普罗旺斯
Completion Date:	2002	完工日期：	2002年重新装潢
Photographer:	Marcel Jolibois	摄影师：	Marcel Jolibois

In the heart of the village of Saint-Rémy-de-Provence, Les Ateliers de l'Image enjoys a superlative geographical situation. The hotel began its history in the building that, until the 1970s, was the old Music Hall cinema. Antoine Godard, a passionate photographer, wanted to create a welcoming place here for a public who appreciated photography. Since 1998 it has been a relaxed luxury hotel, an establishment that cannot be judged using the normal criteria, with 32 rooms and suites in very contemporary design.

The suites are pure and elegant. Perched in a magnificent plane tree is found a "tree-house suite". You approach it via a footbridge. The facade, decorated with a balustrade in transparent glass, allows one to get the full benefit of the landscape. Some rooms have their own hammam, others beautiful teak terraces. The minimalist architecture and interior decoration mix blonde wood and glass, and the tonalities of crimson, cypress or gooseberry. Each room is an exhibition space, displaying the original works of Bernard Faucon, Jean-Pierre Sudre, Romeo di Loreto, Craig Stevens...

The gastronomic restaurant "Le Provence" offers a unique view over the Alpilles. In its light plays an important part, along with attractive furniture. For the interior, a mixture of pale woods and red fabrics, amusing lamps in the form of clouds reinforce the Zen side of the setting. On the terraces, wooden and metal tables, surrounded by laurels and white roses, harmonise with a pool and a stream. There is also the sushi bar "Origami Nouveau", integrating itself well with the natural scene, while the "Resto'bar", in the former cinema, brings its former incarnation back to life with numerous elements: ceiling, seats, projector, photographs...

"影像工作室"酒店位于法国普罗旺斯地区、圣雷米村的中心，具有极佳的地理位置。它所处的建筑物在1970年代以前是该村庄的音乐电影院，热爱摄影的安东尼-高达想重新将这里塑造成一个能够接待爱好这项艺术人士的场所。因此，一个与众不同的豪华休闲旅馆终于在1998年开幕启用，它提供了32个充满当代设计感的套房和一般客房。

酒店套房具有优雅纯净的风格。其中一个树屋套房位于一株壮丽的梧桐树上，借助一座天桥与外界连通；树屋的立面使用了透明的玻璃，让在其中的房客能够尽情观览四周的风景。某些套房拥有私人的蒸汽浴室，其他的套房则拥有原木铺制的阳台。极限主义的建筑与室内设计风格，融合了白色木料和玻璃，带着朱红、柏绿、醋栗等不同的自然色调。这里的每一个房间都是一个展览空间，陈列着贝尔纳-弗孔、让皮埃尔-苏德、罗密欧-迪洛瑞多、克瑞格-史蒂文斯……等著名摄影师的作品。

名为"普罗旺斯"的美食餐厅拥有面对阿尔比勒山区的壮丽视野，此餐厅设计的特色在于大量运用自然光线，并十分讲究家具的陈设。室内装潢采用了浅色的木材与红色的织品，加上云朵造型的灯罩，给予空间一种禅的意境；户外露台上设置了以木材和金属制成的桌椅，它们被月桂树和白蔷薇所围绕着，并与水池和水道产生一种趣味关系。此酒店还拥有一个新的日式寿司餐厅，与一旁的自然景观和谐相映；另外有一个位于原先电影放映室里面的餐厅酒吧空间，设计师采用旧电影院中的天花板、座椅、投射灯、照片……等元素而重新为餐饮空间进行场景设计。

托古洛之家
Troisgros Hotel

Location:	Roanne, France	地点：	法国，罗昂
Completion Date:	2007	完工日期：	2007
Photographer:	Wijane Noree	摄影师：	Wijane Noree

When Michel and Marie-Pierre Troisgros undertook the renovation of this institution, which was formed of the restaurant and the hotel of Roanne, they were in no doubt that they wanted to create a contemporary setting, but, one that was in a certain way timeless. A reputed interior architect gave form to this new setting and marked out its direction. Others followed over the years but each one retained this attention to the light and to tranquility, which is nevertheless crossed by flashes of colour that negate any drowsiness: touches of lemon yellow, aniseed green, Indian pink, which punctuate the sometimes quasi-Japanese spaces.

Marie-Pierre Troisgros has made well-being and elegance evolve in the hotel: 16 light and calm rooms in soft tones, natural materials and with pure forms are arranged around an interior garden. For booklovers, a library richly stocked with culinary and art books has been created. And everywhere paintings by contemporary artists like Favier, Traquandi, Shütte, Tusek, Dörner, or impressionists such as the Roannais painter Jean Puy, emit an unforgettable ambiance. Michel Troisgros says that he draws energy from being around artists: their works are like injections of memory. So much so that a great coherence exists between the general atmosphere, the artworks scattered all around and Michel Troisgros' cuisine: all three speak of simplicity, calm, light and spirit.

当托古洛家的米歇尔以及玛丽皮埃尔在重新整建这个在罗昂地区极具代表性的酒店以及餐厅的时候，他们毫不犹豫地选择了具有当代感的空间风格，但是，就某种程度而言，他们想要的其实是一种超越时间的风格。在一位知名室内设计师的规划下，所有的整建工作都以塑造这个风格为目的。在接下来几年的时间里，接手的设计师们也以同样的意念来继续完成这里对光线和宁静的要求。但为了避免产生令人麻木昏沉的感觉，他们采用了像是柠檬黄、杏仁绿、印度红……等亮丽的颜色，以轻盈点缀的方式来为这里几乎可以说是带有日本风格的静谧空间添增些不同色彩。

玛丽皮埃尔－托古洛试图将酒店朝向创造优雅舒适的环境来发展，因此其围绕着内庭花园而设置的16个客房，个个明亮幽静、色调柔和，并采用自然材料与纯净线条来塑造风格。酒店还特别设立了一个美食图书室以满足爱好烹饪者的欲望。旅馆墙上到处挂着像是法维叶、塔卡迪、苏蒂、屠瑟克、多纳……等当代艺术家的作品，还有罗昂地区的印象派画家让－浦伊的作品，这些作品在空间当中散发着一种令人难忘的氛围。经常与这些艺术家接触、交换心得的米歇尔－托古洛说，这些作品就像能够重新唤醒人心的针剂。在酒店的整体氛围、四处散布的艺术作品以及米歇尔－托古洛精心烹调的佳肴之间存在着一种和谐，所有的一切都带有单纯宁静、灵性明亮的特质。

蒙达龙贝尔酒店
Montalembert Hotel

Designer:	Christian Liaigre
Location:	Paris, France
Completion Date:	2007
Photographer:	Antoine Schramm

设计师：	克里斯提安—里耶格
地点：	法国，巴黎
完工日期：	2007
摄影师：	Antoine Schramm

The Montalembert Hotel is ideally located in the heart of Saint-Germain-des-Prés on the Left Bank. Close to the Orsay and the Louvre museums, this boutique hotel, built in 1926, is the perfect combination between new design and cosy ambiance.

Contemporary or classic, combining Louis-Philippe antiques with a modern twist, each room and suite features innovative bronze lights cast by Eric Schmitt, which illuminate Jean-Pierre Godeaut's photographs and Giuseppe Castiglioni's engravings. All the bathrooms have been designed using chrome and Cascais marble floors. In the very spirit of the French architect Christian Liaigre, the bedspreads bring taupe and tobacco tones to the Montalembert's new cosy beds. The seven refined and elegant junior suites and suites will make people feel really spoiled. On the 8th floor, the mansards bring a typical Parisian touch to the junior suites and suites overlooking Paris and the Eiffel Tower.

Comfortable grey and taupe cushions, beige linens and oak wood add to the elegant feel of the restaurant. A wide range of dishes come under four main categories: earth, sea, vegetable, and sun. For business meetings, teatime or evening drinks, the library and its fireplace offers an intimate and cosy corner. And when spring rolls around, the terrace becomes the ideal place to enjoy lunch or dinner in a peaceful atmosphere…

这个临近罗浮宫与奥塞美术馆、位于巴黎塞纳河左岸圣杰尔曼—德培区的蒙达龙贝尔酒店兴建于1926年，其建筑构思完美结合了创新的设计风格与舒适的空间氛围。

古典或当代，每一个房间或是套房都有着路易—菲立浦带来现代意味的古色家具，艾立克—史密特设计的青铜灯，映照着让皮埃尔—古铎的摄影作品以及季斯培—卡斯蒂里奥尼的版画，所有的客房都铺有卡斯克地区出产的花色大理石地板，另外有建筑师克里斯提安—里耶格所设计的充满法兰西式风格的室内装潢，床组采用了烟草色的毛料，在总计7个套房和小套房里有着细致优雅的空间风格，能让在这里的旅客充分地放松，位于第8层楼斜屋顶下非常法式的套房空间里有着俯视巴黎以及埃菲尔铁塔的迷人景观。

酒店的餐厅在舒适的灰色毛皮枕、羊毛色布料以及橡木材质的装潢下显得更优雅，这里并采用了演伸于土地、海洋、植物以及太阳的4个主题系列的多样化餐具，让在此地进行的商业餐会、下午茶时光以及傍晚小酌都更有滋味。图书室与壁炉空间位于亲切舒适的角落，而当春天来的时候，户外的阳台空间就成为宁静想用午餐或是晚餐的最佳场所……。

史克丽波酒店
Scribe Hotel

		设计师：	贾克-格兰奇
Designer:	Jacques Grange	地点：	法国, 巴黎
Location:	Paris, France	完工日期：	2007
Completion Date:	2007	摄影师：	Fabrice Rambert
Photographer:	Fabrice Rambert		

With its legendary history and an enviable position in the heart of Paris, the Scribe Hote. is the incarnation of Parisian chic. The new decoration of the Scribe was entrusted to Jacques Grange. The Scribe Hotel is invested with the charm of a grand residence. The decoration dares to use precious and exceptional elements. A beautiful harmony of style underlines the unity and identity of the Scribe. This identity cultivates a history, is inspired by the hotel's environment and inscribed in its time.

Chequered stone floors, fluted columns and large crystal chandeliers welcome the visitor at the entrance to the Scribe with a classicism tempered with contemporary materials. In the vast hall where fireplaces, deep velvet sofas and large bookcases offer a cosy comfort, as under the transparent glass ceiling of the Café Lumière, historic references and touches of Art Deco, or even more contemporary, modernity, mix to create a very Parisian style, both of the moment and timeless.

Along the corridors of the different floors, an astonishingly imaginative decor evokes the legend of the Scribe and the important people in its history. In the rooms, suites and duplex, woven horse hair, velvets, brocades and tweeds are mixed. The furniture plays with affectation in the spirit of the Art Deco style, and harmonises with ultra-modern equipment. In its world of opalescent whiteness, the spa is a true bubble of serenity.

拥有传奇性历史、位于人人称羡巴黎市中心的史克丽波酒店是巴黎风华的忠实再现。酒店新的室内设计由设计师贾克-格兰奇负责，企图在此塑造豪门宅第迷人风格，并勇于展现珍奇与独特的性格，同时以风格式样的和谐来强调出酒店整体的个性。这个由所处环境引发灵感而塑造成的性格，造就了一段镶嵌在当代时光里的新历史。

酒店入口的棋盘图案石造地板、肉桂色的柱列和巨型的水晶吊灯，皆采用当代的材料重新诠释、呈现出一种古典主义空间氛围。在宽广的大厅里，壁炉、令人深陷的沙发、巨大的藏书架让此地呈现出柔和温暖的舒适空间氛围，就像罩在玻璃采光顶之下的"光明咖啡厅"的空间气氛一样。此地融合了历史典型、装饰艺术现代风格或是更具当代性格的笔触，成功地塑造出一种深负巴黎韵味的风格，既时尚又永恒。

楼上的走道空间里的奇特装饰风格令人联想到史克丽波酒店的辉煌历史以及其中的传奇人物。在一般客房、高级套房与楼中楼式客房里，皆融合采用了各式毛皮、天鹅绒的织品材料；家具则以装饰艺术的风格来设计，搭配着极具现代感的设备。在白色半透明环境中的水疗养生中心(spa)，就像是一只名实相符的悠然气泡。

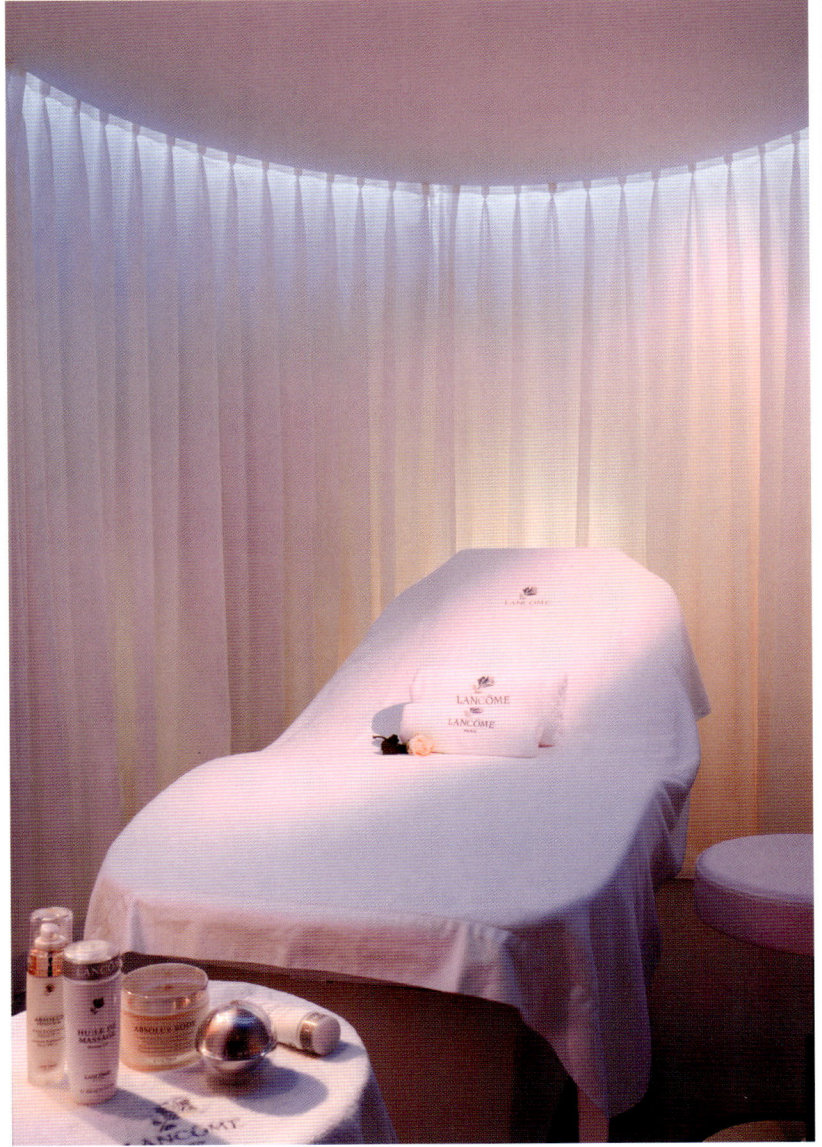

威斯汀酒店
The Westin Paris

Designer:	Pierre-Yves Rochon (building renovation), Jacques García (Le First Restaurant), Sybille de Margerie (rooms)
Location:	Paris, France
Completion Date:	2008
Photographer:	Wijane Noree (pp.188,190,191,193), Eric Cuvillier (pp.189,192)

设计师:	皮埃尔–伊夫–罗逊(建筑整修), 贾克–加西亚(Le First 餐厅), 席比勒–德–玛格丽(客房)
地点:	法国, 巴黎
完工日期:	2008
摄影师:	Wijane Noree (pp.188,190,191,193), Eric Cuvillier (pp.189,192)

Built according to the plans of the architect Henri Blondel, the hotel was inaugurated on the 6th of June 1878 to welcome visitors to the Exposition Universelle. The 47 000 m² building forms a quadrilateral over five floors, bordered by four roads including the rue de Castiglione, where the main entrance of the hotel is found.

Today, the decoration of the establishment's 438 rooms and suites has been given a whole new look by the designer Sybille de Margerie. New life has been breathed into the place using a skilful mix of classicism revisited and contemporary decor, evoking a comfortable Parisian apartment more than a hotel room. In sum, it offers an aesthetic and functional decoration in a classic setting with historical cachet.

The historic public rooms of the Westin Paris are among the most sumptuous salons in the capital. Proof is found in the Imperial Salon with its Napoléon III decor, which has been classed as a Historic Monument since 1972. Its fourteen gilded fluted columns support the ceiling which is lavishly painted and illuminated by three immense chandeliers, flooding the frescoes and silk draperies with light. One side of the salon houses a monumental chimneypiece clad in marble and supported by bronze caryatids. An imposing clock dating from 1878 ornaments the mantelpiece.

Situated on the ground floor is The First, a Parisian boudoir restaurant designed by Jacques García. The first of its kind in the capital, The First offers a contemporary interpretation of brasserie cuisine. With its refined mix of purples, velvets and silks, The First is a unique place.

由建筑师亨利–布龙岱尔设计、在1878年6月6日开幕启用的威斯汀酒店，原来是为来巴黎参观世界博览会的游客所建造的，方形的建筑平面，楼高5层，一共47000平方米的面积，酒店四边临路，自成一个街坊，主要的入口位于伽斯底里内路上。

酒店现今的438个标准客房以及豪华套房，在经过设计师席比勒–德–玛格丽的重新设计之后令人耳目一新。设计师微巧地融合了重新诠释的古典主义以及具现代精神的装饰风格，她企图在此创造的不是一间间的旅馆房间，而是重新营造出一个舒适的巴黎公寓空间。总体而言，威斯汀酒店的室内设计是在其历史性古典的外貌之下，将美学与实用性相结合。

威斯汀酒店的各个沙龙、大厅是在巴黎的历史建筑中最富丽堂皇的作品之一，其中有在1972年被列为国家历史遗产来保护的拿破仑三世风格的帝国厅，耸立在其中的14根镀金的圆柱撑托着其上的彩绘壁画天花，从天花板垂挂而下的三盏巨大水晶吊灯漫射散发的光线，将四周的壁画以及丝织帘幕溶入在一种特殊的氛围中。大厅中还有一座由青铜女像柱所撑托着的宏伟大理石壁炉，其上放置的一座1878年大钟让空间更显华贵。

位于地面层、由贾克–加西亚所设计的第一餐厅，以小客厅形式呈现，是巴黎同类型餐厅的创始者，在细致结合了丝料和绒布的两种紫红色彩的空间中，供应法国酒馆现代风格的料理，使得第一餐厅成为巴黎一个十分独特的场所。

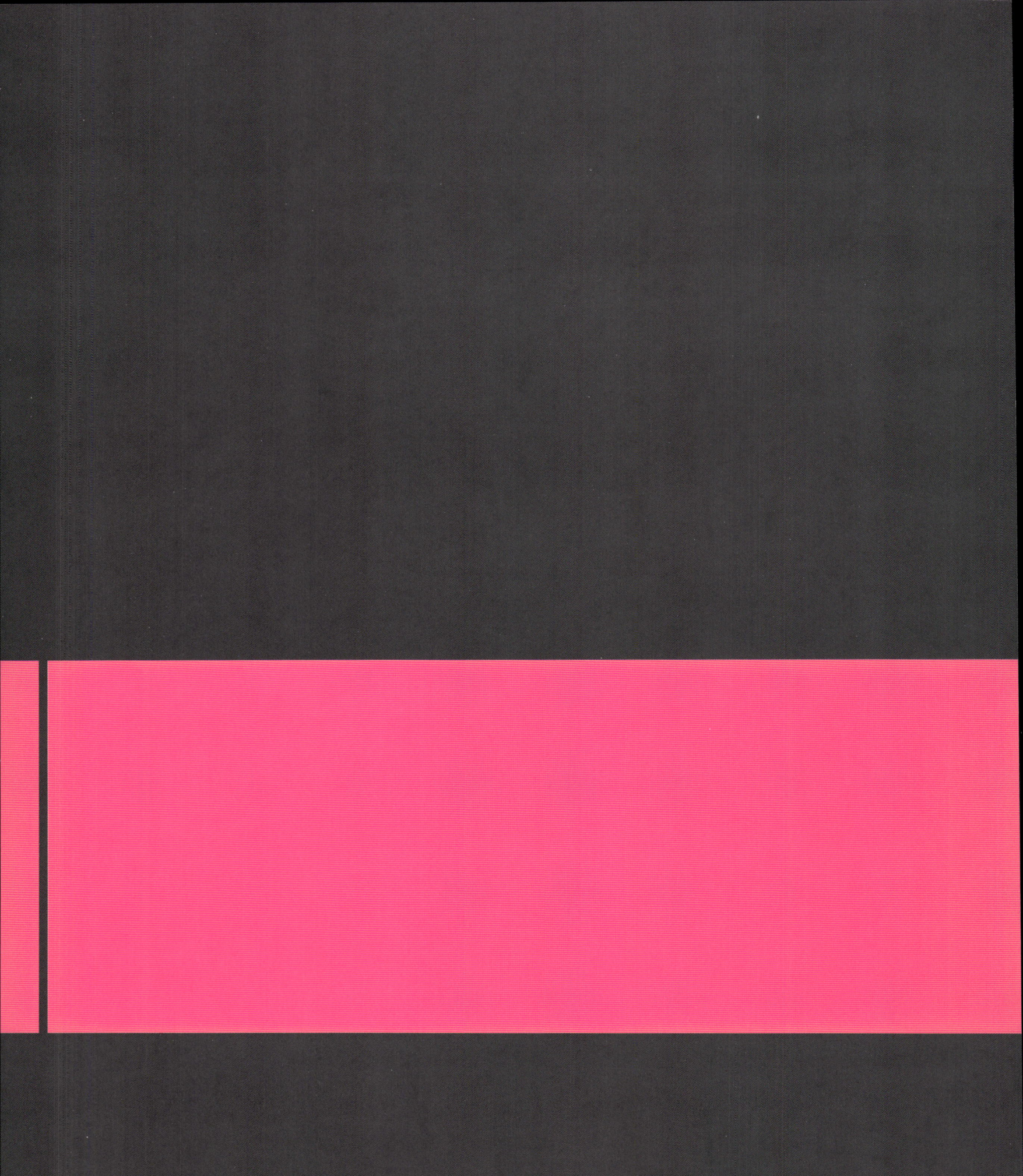

Luxury - Classicism - Elegance

Luxury Palace Hotels

豪华酒店

华丽–古典–高雅

雅典娜酒店

Plaza Athénée Hotel

Designer:	Patrick Jouin (restaurant-bar)
Location:	Paris, France
Completion Date:	2001 (restaurant-bar), 2007 (various)
Photographer:	Thomas Dhellemmes (p.196), Philippe Derouet (p.197), Plaza Athénée (pp.198-120), Christoph Kicherer (p.121)

设计师：	派崔克-乔恩 (餐厅-酒吧)
地点：	法国，巴黎
完工日期：	2001 (餐厅-酒吧)，2007 (多处空间重新装修)
摄影师：	Thomas Dhellemmes (p.196), Philippe Derouet (p.197), Plaza Athénée (pp.198-120), Christoph Kicherer (p.121)

In 1852, the Emperor Napoléon III commissioned Georges Haussmann to transform Paris into a modern and bustling capital that would enjoy worldwide recognition. During this period, many unique hotels were built which have today become symbols of traditional French style. In 1911, at 25 Avenue Montaigne, the Plaza Athénée Hotel opened at the same time as the nearby Théâtre des Champs-Élysées: the hotel soon became the gathering place for the most prominent composers and performers of the day. By 1936, the hotel had achieved international fame. The Galerie des Gobelins and the restaurant Le Relais Plaza, its salons and apartments hosted many great personalities of this period. When Christian Dior opened on Avenue Montaigne in close proximity to the Plaza Athénée, a new era began. Avenue Montaigne quickly turned into the avenue of haute couture, and the Plaza Athénée became Paris' focus for fashion and the arts.

All 188 bedrooms and suites of the Plaza Athénée were entirely refurbished in 1999. The bedrooms and suites on the first six floors are decorated in French classical style, furnished in the period of Louis XV, Louis XVI and Regency, with fabrics copied from original designs, authentic mouldings and fireplaces. On the 8th floor, the unique Royal Suite and Terrace Eiffel Suite, true Parisian apartments, with classic style for one and Art Deco for the other, offer a staggering and incomparable view of the majestic Eiffel Tower. Throughout the hotel, sketches from well-known fashion designers adorn the corridors and suites. State-of-the-art technology has been installed to provide maximum comfort and convenience throughout.

法国欧斯曼男爵在遵循拿破仑三世的意旨之下，于1852年开始进行巴黎的都市更新工程，要将这里改造成举世公认、充满活力的现代化首都。许多在这一段时间建造的酒店，今天都成为法兰西风格的象征。1911年，位于蒙田大道25号的雅典娜酒店和旁边的香榭丽舍剧院同时开幕，并且很快地就成为当时活跃的作曲家以及表演者的群聚之地。1936年雅典娜酒店享有国际知名的声誉，酒店的葛伯兰廊厅、"广场驿站"餐厅、沙龙与客房，吸引了许多知名的人士在此居住停留。直到克里斯汀-迪奥同样选择在本区开店之后，一个新的时代从此展开，蒙田大道因而成为时尚界的大道，而雅典娜酒店也成为巴黎流行与艺术的焦点。

合计188间的一般客房与豪华套房在1999年全面更新，从二楼到七楼的客房以法兰西古典风格来装修，搭配着路易十五、路易十六时代式样的家具，并且采用复制原设计图案的布料、纯正的装饰线脚与壁炉。位于九楼皇家套房和艾菲尔露台套房是名副其实的巴黎式公寓，分别以古典风格和新艺术风格来装修，并且拥有令人迷醉、坐揽艾菲尔铁塔的宏伟景观。整个酒店的走道和客房都由知名的设计师来设计，结合了最先进的科技设备来提供最舒适合宜的服务。

At the Plaza Athénée, the pleasures of the palace are celebrated, cultivated, brought to life and expressed in different venues, each in its own way combining the art of living with delicious cuisine. The gastronomic Alain Ducasse at Plaza Athénée reflects the art of fine dining in the style of Louis XV. During the summer months La Cour Jardin, probably the most famous courtyard in Paris opens with its distinctive red parasols and colourful foliage. The Plaza Athénée is also famous for its unique and contemporary bar designed by Patrick Jouin in 2001. After almost a century, the Plaza Athénée continues to be the top choice of celebrities as the place to stay in Paris.

雅典娜酒店提供的雅致皇宫式的享受与愉悦充分显现在各个层面，并结合了生活艺术和独特美食。由美食大厨亚兰-杜卡斯所主持的餐厅以路易十五的风格来展现精致的饮食艺术。名闻遐迩的酒店花园中庭，在夏日时节以其大红色遮阳篷和色彩缤纷的植栽来迎接嘉宾。酒店的另一个著名场所是由设计师派崔克-乔恩在2001年所设计、极具当代风格的独特酒吧。一个世纪以来，雅典娜酒店一直是举世知名人士停留巴黎时最佳的地点。

巴尔扎克酒店
Balzac Hotel

Designer:	Anne-Marie Sabatier
Location:	Paris, France
Completion Date:	2007
Photographer:	Janos Grapow

设计师：	安妮玛丽-萨巴蒂埃
地点：	法国，巴黎
完工日期：	2007
摄影师：	Janos Grapow

Dominating the rue Balzac, the hotel was originally built for the director of the Paris Opera. It enjoys a particularly quiet and peaceful location, and clearly lives up to its credentials as a luxury hotel while, at the same time, being one of the most romantic hotels in Paris.

Anne-Marie Sabatier, who formerly worked with Pierre-Yves Rochon, has created her own take on the neo-classical style, with a blend of different styles and periods. She has turned towards rich vibrant colours in her choice of fine fabrics: turquoise blue, chestnut, plum and copper for the bedrooms and public spaces. The Italian furniture for the suites and bedrooms, and the panelling in the hall and the lounge, is classic in style with a touch of the contemporary. The lamps are produced in a small family factory in Murano. The carpets and rugs stylishly incorporate the Balzac Hotel's own monogram. There are extremely original 18th and 19th century lithographs digitally layered and coloured, depicting scenes of life in Paris at that time.

Situated on the top floor, the Royal Suite offers a stunning view of the whole city from its terrace. Next to it, the Presidential Suite is distinguished by the warmth and intimacy of its decor. The other 11 suites consist of Junior Suites and Corner Suites. Today's travellers will appreciate the 57 luxury rooms, equipped as they are with state-of-the-art technology in a luxurious environment.

There is a magnificent glass roof nestled in the heart of the hotel. A muted lounge is home to the Lounge Bar, with its subtle palette of sensuous tones. 3 VIP Meeting Rooms are available for use by guests wishing to meet in private. Balzac Hotel invites its guests to share the excellence of the French "art of living".

该酒店位于巴黎的巴尔扎克路上，享有极为宁静的四周环境，并散发出豪华酒店的气质，同时也是巴黎最浪漫的旅馆之一。

酒店的设计师安妮玛丽-萨巴蒂埃曾经是名设计师皮埃尔伊夫-罗逊的合作伙伴，她在此创造出的新古典主义的氛围，呈现出多个时代与风格的对照。她在客房与公共空间使用了亮丽的颜色来凸显酒店的珍贵织品：土耳其蓝、冰栗褐、枣紫红和铜黄。套房与一般客房皆采用了意大利式的家具，大厅与吧台交谊厅的细木装饰在古典的风格下带有一丝现代的色彩，灯具由意大利最具盛名的玻璃工业品产地穆拉诺的一家家族式工厂负责制造，地毯则由带有巴尔扎克字母的花纹精心织成。墙上以数码科技投射的 18 和 19 世纪的非常有创意的石板画，展现着巴黎美好时代的生活场景。

位于顶楼的皇家套房透过其户外露台拥有观赏巴黎的不凡视野，与其比邻的总统套房则以具私密性的温暖色调在装饰上取胜，另外 11 个小型套房以及角间套房也都各具特色。此外，57 个高级客房提供了住客优雅的空间装潢以及先进的科技设备。

酒店中庭的玻璃采光罩美轮美奂精致，并为建筑物内部带来明亮的光线。使用了感性色泽来装饰的沙龙酒吧空间散发着温柔的气息，另外有 3 间贵宾沙龙提供作为私人聚会的用途。巴尔扎克酒店透过这些空间来让旅客尽情分享巴黎生活艺术的精髓。

布里斯托酒店
Le Bristol Hotel

Location:	Paris, France	地点：	法国，巴黎
Completion Date:	2007	完工日期：	2007
Photographer:	Guillaume de Laubier	摄影师：	Guillaume de Laubier

Resonating with Parisian life and culture, Le Bristol Hotel is an exceptional property that has transformed service into an artform. With its vast classical French garden ; a true heaven in the centre of the city, Le Bristol Hotel is a unique and rare destination in itself. Located a stone's-throw away from the Champs Elysées, the hotel is situated in the Faubourg Saint-Honoré, with luxury antique dealers and art galleries just moments away.

Each day of the week the two Michelin-starred chef Éric Frechon creates an exceptional gourmet cuisine that is both lavish and innovative. The elegant bar welcomes a Parisian and International clientele for afternoon tea, lunch and dinner. The legendary Fashion Teas combine runway shows from the big names of French and International couture with a refined English-style tea. The 161 rooms and suites portray a luminous beauty in softly nuanced shades with rare woods and precious fabrics, punctuated by Louis XVI furniture and antique furnishings.

Located on the 6th floor, the swimming pool is surrounded by teak and glass, opening up onto a large sundeck with commanding views of Paris. A fitness room, with the latest equipment, and an Anne Semonin Spa complete this haven of luxury. The opening of Résidence Matignon at the beginning of 2009 will provide 22 additional rooms and 4 suites ; some with magnificent views of the Eiffel Tower, in addition to a grill-restaurant seating 100 ; all directly accessible from the rue du Faubourg Saint-Honoré.

位于佛布－圣欧诺雷大街上的布里斯托酒店，近邻香榭丽舍大道，以及精品名店、古董商店和艺廊集中的地区。酒店拥有宽广的法式庭园，是位于巴黎城市中心的一个平静避风港，也是巴黎一个罕见独特的场所。它反映了巴黎的生活与文化，并将服务业转化为艺术的展现。

拥有两星评鉴的名厨艾立克－弗雷雄在此每天提供风味出色、气派而有创意的美食，优雅宁静的酒吧餐厅提供巴黎过往旅客喝下午茶以及正餐的服务。在名为"风尚名茶"的茶馆里可见的是出入往来的法国和世界时尚界名人，在此地品尝英国茶的细致风格。这里的161间客房以及73间套房都拥有一种亮丽美感、温和变化的色调、罕见精致木质装修、珍贵的织品布料，搭配着路易十六风格的家具以及著名画家的作品。

在酒店顶楼，有以玻璃和木材装修的游泳池，伴随着宽敞、可以俯瞰拥抱巴黎景观的日光浴室。还有桑拿、设备齐全的健康养生中心以及养颜美容中心，使酒店的服务更臻完美。预计在2009年年初开幕的马蒂侬宅邸，将为酒店增加22间客房以及4间套房，其中的一些房间拥有面对埃菲尔铁塔的绝佳视野；另外还有一间可容纳100人的烧烤餐厅以及一个酒吧，这些两个空间都可以直接从佛布－圣欧诺雷大街上进入。

克莱耶尔城堡酒店
Château Les Crayères

Designer:	Pierre-Yves Rochon	设计师：	皮埃尔伊夫–罗逊
Location:	Reims, France	地点：	法国, 兰斯
Completion Date:	2007	完工日期：	2007
Photographer:	Philippe Chancel	摄影师：	Philippe Chancel

The Château Les Crayères belongs to Reims and to its region: from its construction, in 1904, onwards, it has been a privileged and elegant gathering place for the city and the world of champagne, a tradition that has never waned. When the château opened its doors as a hotel in 1983, Les Crayères also became a gastronomic meeting place.

Pierre-Yves Rochon has been linked to Les Crayères for almost 25 years. He recently completed suite n°20, which prefigures the spirit that he wishes to breath into this hotel: modernity, integrating all the new technologies and services that each client has the right to expect from such a hotel, while preserving the place's unique spirit. "My challenge is to achieve a perfect fusion between the past and the future through elegance", he likes to explain. Through his international experience and the numerous hotel renovations that he has managed, Pierre-Yves Rochon has a very polished perception of what a luxury hotel should be today. Palaces and "executive hotels" are no longer enough: one has to create places for clients who have time with hotels that have "a soul".

The kitchen is in the hands of the young and talented chef Didier Elena. After four years spent in New York heading up the kitchen of the Alain Ducasse restaurant Essex House, Didier Elena has breathed new life into the precious gastronomic heritage of the restaurant of Château Les Crayères.

位于兰斯香槟区的克莱耶尔城堡建造于1904年，从那时候起，它就成为是兰斯市和香槟世界里的一个优雅华贵的场所，而这个光荣的传统至今始终未有改变。城堡于1983年成为旅游酒店对外开放，从此亦成为人们品尝美食的相约之处。

在设计师皮埃尔伊夫–罗逊和这个酒店长达25年的合作关系之下，他新近完成的20号豪华套房里预先展现了其个人所希望在此塑造的新风格：不仅具有符合现代旅客所期待的最新科技与服务的当代精神，同时也保存了城堡的场所精神。设计师说："在这个设计方案里，最大的挑战来自于如何以高雅的手法让过去和未来完美地结合"。拥有非常丰富的国际经验、经手众多的酒店设计方案，皮埃尔伊夫–罗逊对于今日豪华酒店应具有的面貌有着非常深刻精辟的见解。宫殿式酒店以及贵宾酒店已经不足以满足客户的需求，他说："我们要为'有时间'的旅客消费者塑造新的场所，现今必须营造的是有'灵魂'的酒店"。

这里的美食供应由年轻有才华的主厨帝迪耶–艾连纳负责，他过去在纽约阿伦–杜卡斯的爱瑟斯之家餐厅的4年里位居重职，如今在这里重新为克莱耶尔城堡酒店的宝贵美食传统开辟新方向。

乔治五世酒店
Four Seasons - George V Paris

Designer:	Pierre-Yves Rochon	设计师：	皮埃尔伊夫–罗逊
Location:	Paris, France	地点：	法国，巴黎
Completion Date:	1999	完工日期：	1999
Photographer:	George V Hotel	摄影师：	George V Hotel

Two experts are responsible for the renaissance of the George V. Richard Martinet, in charge of the renovation of the facade and the exterior buildings, chose continuity and has given the original building back its brilliance as faithfully as possible. Inside, Pierre-Yves Rochon has made the Four Seasons Hotel George V one of the most beautiful palaces in Paris, integrating technology into the decoration using the utmost discretion.

The entrance hall reflects the mixture of French and English influences that one finds in the rest of the hotel. On the floor, marble is inlaid in a complex marquetry of Sienna yellow, grey and beige tones. To the left of the lobby, one finds in the English Salon with its large bookcases in sycamore wood. The draped curtains, in antelope-coloured velvet, open onto the interior courtyard of the hotel. The Bar, in warm tones, offers a refined setting that extends along the gallery which houses a splendid Savonneries carpet 20 metres long. At the end of the gallery on the left, two wrought iron doors open onto the dining room of Le Cinq restaurant, vast and luminous in tones of gold and grey. Two large Regency mirrors in sculpted, gilded wood give depth to the space and reflect the garden of the Marble Courtyard.

乔治五世酒店的整建工程由理查–马逊内以及皮埃尔伊夫–罗逊两位设计专家来进行。理查–马逊内负责建筑的外观立面，选择了以忠实延续建筑物固有的风格来重新赋予酒店新生命；皮埃尔伊夫–罗逊则以严谨审慎的鉴别力将高科技与室内装饰结合，使乔治五世酒店成为巴黎最美丽的高级酒店之一。

贯穿整个酒店室内空间、同时融合了英法两国精神的设计风格在酒店大厅里便一览无遗，大厅地面则采用了由意大利黄色、灰色以及米白三种颜色精心组合的大理石地板。大厅右侧的英格兰沙龙有着原木制成的宏伟藏书架，其旁的落地窗垂挂着大面绒布帘幕，并向着酒店的内部中庭展开。采用暖色调、带有纯净简洁风格的吧台空间，一直延伸到一个悬挂着一幅长达20米织锦壁毯的长廊空间，长廊底端左边的两扇生铁铸成的大门则带领人前往宽敞明亮、以灰金双色为主色的乐桑客餐厅。餐厅里两面以镀金木雕为框、具摄政王时代风格的大镜不仅赋予了室内空间的深度，也映照着外面名为"大理石庭"的花园。

In the rooms and suites, blues, greens or yellows blend into an ivory-dominated whole. Tone on tone is very much the theme here. Painted furniture patinated in antique style recalls the colours of the fabrics in the rooms. In a total of 61 sober and elegant suites, Pierre-Yves Rochon has created harmonies of straw yellow and gold, sky blue and sapphire, or sea and emerald green with silks and damasks from Lyon. Whether one focuses on an Empire or Presidential suite, paintings from the French schools of the 18th and 19th century, the chic spirit of Parisian interiors, the terraces with superb views or a deep sofa, the atmosphere in each of the rooms is subtle and different.

In the Spa a trompe-l'œil mural invites bathers to plunge into the wide avenue leading to a château and its large lake. The ceiling represents a blue sky and the balustrade in stone gives the impression of being in the open air.

所有客房与豪华套房皆以象牙白为主色调，搭配或蓝、或绿、或黄的颜色。客房内的仿古家具与其织布颜色相呼应。在61个简雅高贵的一般套房里，设计师运用了稻草黄、金黄、蔚蓝、宝石蓝、湖水绿、翡翠绿的丝质毛料以及印染织布；在帝国套房和总统套房则吊挂着18和19世纪法国学院派的绘画、搭配着巴黎风格的室内装修、拥有不凡的祝野以及令人深陷的沙发，每一间套房的空间气氛都非常雅致并且各不相同。

在酒店的养生水疗中心(spa)，墙上几可乱真的壁画邀请人们进入一条宽敞的廊道、前往尽头的城堡与水池，其天花板表现着湛蓝的天空，石砌栏杆让人仿佛置身室外。

洛岱乐酒店
L'Hôtel

Designer:	Jacques Garcia	设计师：	贾克-加西亚
Location:	Paris, France	地点：	法国 巴黎
Completion Date:	2006	完工日期：	2006
Photographer:	Anne-Laure Jacquart	摄影师：	Anne-Laure Jacquart

234+

L'Hôtel situated on Paris' Left Bank, which was relaunched in December 2006 following a large-scale renovation, has captured the quintessence of English chic and refinement. Nestled in the heart of literary and artistic Paris, L'Hôtel is surrounded by beautiful art galleries, in the shadow of the dome of the Académie Française, the Académie des Beaux-Arts and the Louvre.

It was naturally to one of the world's ambassadors of luxury, Jacques Garcia, that L'Hôtel turned once again to give the establishment a new look, and to create a more sensual and relaxed feel throughout. A legendary Parisian address in the past, L'Hôtel rapidly acquired prestigious guests including the writer Oscar Wilde, who died here, as well as Jorge Luis Borges, Ava Gardner, Frank Sinatra, and Dame Elisabeth Taylor with Richard Burton...

Situated in the heart of the Saint-Germain-des-Prés of Jean-Paul Sartre and Simone de Beauvoir, Ernest Hemingway and Oscar Wilde. L'Hôtel has quietly cultivated an exceptional tradition of hospitality where refinement and history have always gone hand in hand. Its 20 rooms and suites, revisited as cinematic memories with their views of the rooftops of Paris, include the Art Deco Mistinguett room with furniture from the music hall star's own bedroom ; the Oscar Wilde room with its huge, emerald, peacock-design mural inspired by an engraving of Wilde's dining room in London, and the large suite on the top floor, with its superlative view of the churches of St Germain and St Sulpice.

洛岱乐酒店位于巴黎塞纳河左岸，在经过大规模的整修之后，于2006年年底重新开幕，展现出一种无法归类的高雅细致的英格兰式风格。酒店位于巴黎的文学与艺术中心区域，被许多美丽的艺廊所环绕，并且与法兰西学院、巴黎美术学院以及罗浮宫比邻而居。

在世界知名、擅长豪华风格的空间设计师贾克-加西亚的巧手之下，让洛岱乐酒店重现其空间的价值和风华，成功地塑造出一种魅人、舒适的空间氛围。这里是老巴黎的传奇场所，吸引了许多著名的文学家在此落脚：英国著名的作家王尔德选择此地作为他临终之前的住所，此外还有作家乔治路易斯-博尔赫斯，艾娃-嘉德娜，歌手法兰克-辛纳区，以及影星伊丽莎白-泰勒偕同其夫婿也是著名演员的理查德-伯顿……

酒店所在的圣杰尔曼-德培区的中心曾经是沙特、波娃，海明威与王尔德在巴黎的生活圈，因此酒店在此以毫不虚浮的方式延续着融合了细致品位与历史感的待客传统。可以俯瞰巴黎街景、拥有20间客房与套房的洛岱乐酒店已经成为人们见证巴黎历史的地点。这里有以装饰艺术风格设计的、法国著名歌手密斯丹格苔的套房和其卧房的家具，以及王尔德套房里和他的伦敦住所餐厅一样的大幅祖母绿凤凰壁画，加上窗外面对圣杰曼教堂以及圣叙尔皮斯教堂的优美景观。

Comfort - Elegance - Intimacy

Modern Mansion Hotels

府邸风格酒店

舒适–优雅–亲切

佛朗索瓦一世酒店
François 1er Hotel

设计师：	皮埃尔伊夫–罗逊
Designer:	Pierre-Yves Rochon
Location:	Paris, France
地点：	法国，巴黎
Completion Date:	2000
完工日期：	2000
Photographer:	Fabrice Rambert
摄影师：	Fabrice Rambert

Designed by Pierre-Yves Rochon, the decor of the François 1er Hotel plays knowingly with the contrast between styles in a near-Renaissance spirit and those associated with a completely modern spirit. The volumes are intimate and space is suggested, and maximised, by the lighting.

The spirit of a "private home" is perceived as soon as one enters the entrance hall, where the reception is placed in a small salon framed by columns and curtains, in the middle of which a desk and its armchairs have pride of place. Everywhere, colours and materials have been chosen to create a warm and cosy atmosphere. These harmonies of yellow and red, this mixture of styles, the use, for example, of toile de Jouy, velvet or marble, patterns linked to stripes or to the uniting of long spaces, give the 40 rooms offered by the François 1er Hotel a delicate appeal, just like little jewels. The decoration of the two suites also combines references to the past and today's materials, resulting in a very contemporary comfort.

A winter garden, a restaurant, a tea salon and the conference rooms "Magellan" or "Elysées", which can be reserved for study days or small residential seminars, complete this whole, where the art of living is shown to its best advantage.

由皮埃尔伊夫–罗逊所设计的佛朗索瓦一世酒店展现出一种介于文艺复兴与现代精神之间的装饰风格。酒店空间的体量配置方式趋向于塑造私密性氛围，光线则进一步为空间带来特殊的质量。

自大厅开始，人们便可感受到一种"豪华宅第"的气派，由廊柱与帷幔环绕构成的接待处犹如小型沙龙，中间俨然端置着一个办公桌和若干扶手座椅。空间里的所有颜色与材质的选择，都是为了营造一种热情而柔软的氛围。朱黄相融、风格混搭，使用祝伊花布、绒布或大理石、选取条纹状或可极致重复的图案，种种元素的搭配将酒店40个客房塑造成十分精致、犹如珠宝般的空间。两间套房当中的装饰同样结合了传统与现代的双重设计风格，寻求符合当今舒适需求的条件。

名为"马吉兰"和"爱丽舍"的会议室提供酒店宾客举办研讨会使用，此外酒店还有一个温室、一个餐厅和一个饮茶沙龙，共同将此地塑造成一个充满生活艺术的场所。

夏托布里昂酒店
Chateaubriand Hotel

Architect:	Élodie Sire	建筑师：	爱洛蒂-希赫
Designer:	Philippe Daraux	设计师：	菲立浦-达罗
Location:	Paris, France	地点：	法国, 巴黎
Completion Date:	2007	完工日期：	2007
Photographer:	Sylvain Beche	摄影师：	Sylvain Beche

In the Chateaubriand Hotel, refinement is everywhere, yet deployed with discretion, as an echo to comfort. Enter an atmosphere with a romantic touch: relax amidst calm and harmonious colours. The Chateaubriand Hotel reconciles the discreet luxury of the 19th century with the facilities of the 21st.

Each of the twenty eight rooms is unique ! With the charm of former times mixed with modern comforts, every room has its own personality: designed and equipped in an individual style for the satisfaction of the clients. Everything was designed to achieve a maximum of well-being: precious silk, velvet and satin fabrics and antiques were selected with the utmost care by the owners. Italian marble and precious woods in the bathrooms offer the highest comfort. Natural light flows through the spacious rooms, with windows looking on to the romantic patio or the quietest street off the Champs Élysées, rue Chateaubriand.

The lobby and the reception rooms offer a surprising modernity through the colours of the walls, which are both sober and original, and the diversion of old objects finding new uses, or not finding them at all but simply creating the charm of these welcoming spaces. All this is the successful work of three architects, three decorators and four artiquarians who travelled the world to make the Chateaubriand Hotel look its best, accompanied by the input of the owner himself !

夏托布里昂酒店以婉约的方式将精致品位推至极致，也十分着重空间使用的舒适性。它的室内装饰以浪漫风格为主调，并且到处呈现着宁静氛围与和谐色彩。这里结合了19世纪节制谨慎的华丽风格以及21世纪的舒适便利性。

酒店28间客房的设计，彼此存在着某种和谐关系却互不相似：结合了现今的舒适性和历史的迷人魅力，每一个房间都有其独特的个性，而且都以不同的风格来构思、搭配不同的家具，提供身在其中的旅客最高的舒适感。为了达到此目的，这里大量采用了丝、绒、缎……等珍贵的布料织品来搭配其中的古董家具；在浴室里，运用大理石来搭配罕见的原木；每个房间皆十分宽敞，并且窗户皆面对着酒店的中庭或是香榭丽舍街区最宁静的夏托布里昂路，因此拥有充沛的自然光线。

酒店的大厅和接待沙龙呈现出令人震撼的现代感，其墙面色彩同时兼具婉约与创意，古物新用带来趣味，或者发挥其纯粹的装饰魅力为空间带来迷人氛围。这一切都是结合了3位建筑师、3位室内设计师、4位走遍世界的古董收藏家的贡献，加上酒店主人的用心经营，使酒店设计成为一个成功不凡的案例。

香榭丽舍玛丽农酒店
Marignan Champs-Élysées Hotel

Designer:	Olivier Gagnère		设计师：	奥利维耶-加尼尔
Location:	Paris, France		地点：	法国，巴黎
Completion Date:	2007		完工日期：	2007
Photographer:	Philippe Schaff		摄影师：	Philippe Schaff

Entering the Marignan Hotel, one has the impression of being in a hotel unlike any other... Olivier Gagnère has made this hostelry into a very elegant place where tonalities both feminine (superb mirrors and wall lights, ivory tones and majestic chandeliers) and masculine (leather panels and consoles in dark wood) mix harmoniously. The hotel reception is at once an intimate boudoir and a theatre of Parisian life. It is a charming take on luxury in a warm and welcoming atmosphere.

Two magnificent chandeliers with subtle harmonies illuminate the main lounge, emerging from a cylinder lined with gold leaf. Made by the house of Véronèse, chandeliers, lamps and wall lights were hand-blown in Murano according to the original designs of Olivier Gagnère. The layout of the space offers lots of comfort to its guests, who can choose between the armchairs and sofas pleasantly arranged in the rotonda of the main lounge, or in the alcoves of the small lounge that provides a more intimate and very unique atmosphere.

Each one with a different decor, but faithful to the same idea, the superior and de luxe rooms or duplex suites combine French refinement and an English-style "cocooning" atmosphere. Fabrics with flowery or colonial motifs, ivory-coloured bedlinen, velvet eiderdowns, acacia wood furniture, large sofas... The duplex suites have two independant spaces with a real work area separate from the private space. Certain rooms offer a glorious view over the Butte Montmartre and the Sacré-Cœur. Others even reserve the exclusive privilege of a private terrace where one discovers a unique view of the Eiffel Tower.

一走进位于巴黎香榭丽舍大道区的玛丽农酒店，人们立刻感受到一种与一般旅馆大不相同的氛围……。设计师奥利维耶-加尼尔，将酒店的接待空间塑造成一个高雅的场所，让女性柔美(精美的大镜与壁灯、象牙色调和壮丽的吊灯)以及男性阳刚(皮质面板以及深色原木托架)的特质在此和谐交融。此接待区不仅具备了私密会客厅的空间品质，同时成为反映巴黎社交生活的真实场景，在热情的氛围当中展现奢华的迷人面貌。

大厅中两盏细致和谐的华美吊灯自镶着金叶的珠宝盒伸出，照亮着整个空间。这些吊灯以及其他立灯和壁灯都是奥利维尔-加尼尔的精心作品，并同样由意大利威罗尼品牌出产、在穆拉诺以手工吹制而成。这里有陈设着轻松宜人的沙发和长椅的大圆厅，也有适合亲密谈心的壁龛式座位的小偏厅，酒店的宾客可以自由选择适合其需求的空间。

客房的装饰各不相同，但却带有一致的精神：从高级客房、豪华客房以及楼中楼套房，都结合了法兰西式的细致风格与英格兰式的柔软舒适氛围。花草叶文或是殖民地风格的织品布料、象牙色的绒布床罩与床单、胡桃木家具加上大气的沙发……。楼中楼套房由两个完全独立的空间组成，因此得以设置一个与休息空间完全分开的工作沙龙。有一些房间拥有坐揽蒙马特山丘或是圣心堂的优美景观，另外有几个顶级的房间拥有独一无二面对埃菲尔铁塔景观的私人阳台。

Also designed by Olivier Gagnère, the 15 Cent 15 lounge bar opens up like a jewel box, in an extension of the hotel reception area. It is a precious and elegant place, both contemporary and comfortable, graphic and glamorous, in combinations of purple and fern green, grey tones and striped fabrics. Despite the density of this decor, it gives a feeling of softness.

The lounge bar of the Marignan Hotel is a natural complement to its restaurant Spoon Food & Wine, a coveted and elegant address. Here Alain Ducasse wanted to offer a cuisine inspired by the tastes and flavours of the world. Since September 2005, the restaurant has been dressed in new colours. The architect and interior decorator Jean-François Auboiron has used small touches little by little to make the original decor of Spoon evolve. The result ? More luminosity, more contrasts, more comfort...

一个同样由奥利维尔－加尼尔设计、像是珠宝盒般的 15 Cent 15 交谊酒吧是酒店大厅的延伸空间，这是一个罕见、优雅、现代、舒适而且意象迷人的场所，融合了紫红、翠绿、银灰以及条纹式的各种色彩与装饰，尽管显得丰盛密集，却仍散发着温和柔美的感觉。

酒店的交谊酒吧和它的餐厅"酒食之匙"(Spoon Food & Wine)相得益彰，是巴黎一处难得的雅致场所，名厨亚兰－杜卡斯希望在此推出世界风格的美食。从2005年9月开始，餐厅逐渐改换面貌，经过身兼建筑师与室内设计师的让佛朗索瓦－欧伯龙的设计装潢之后，餐厅变得更明亮，对比性更强，也更为舒适宜人。

丹尼尔酒店
Daniel Hotel

Designer:	Tarfa Salam	设计师：	塔法－莎兰
Location:	Paris, France	地点：	法国，巴黎
Completion Date:	2005	完工日期：	2005
Photographer:	Daniel Hotel	摄影师：	Daniel Hotel

Between the Faubourg Saint-Honoré and the Champs-Élysées, the Daniel Hotel is an "invitation to travel", a place full of references where East and West mingle in the dazzle of the colours, the richness of the materials. The decorator, Tarfa Salam, has followed the Silk Route in the way that it influenced the French style of the 18th century, while the Daniel's owners sourced furniture and objects from London to Washington, from Egypt to Lebanon, in the spirit of a "travel journal".

Green is the colour of the hotel, declined in nuances and various patterns, accompanied by mauve, grey, dusky rose and red. These colours are highlighted by silver and gold. The wallpaper of the salon and the lift was designed in China ; the sofas and armchairs were made in England and the ground floor tables bought in Beirut. The designer Nada Debs designed the bar in ebony and mother-of-pearl. It is placed beneath a glass partition incrusted with feathers and sequins, contrasting with the gold leaf that covers the window or decorates the multicoloured glass roof. The fabrics are equally precious: silks, brocade velvets, satins on the seats, and a multitude of cushions...

In the restaurant, the menu is international, a subtle mix of French, Mediterranean and Asian cuisine in a spirit of cosmopolitanism that goes perfectly with its decoration. As Tarfa Salam says: "I wanted to create a feeling of well-being where each guest feels at home and recognises elements of their own culture during a stay in Paris."

位于巴黎佛布－圣欧诺雷大街和香榭丽舍大道之间的丹尼尔酒店，以它兼容的东西风采著称，透过多样的色彩和丰富的材质，邀请人们进入旖旎之旅。设计师塔法－莎兰以丝路的历程为设计灵感，将18世纪的法国风格重新诠释，并利用酒店主人们自世界各地（从伦敦到华盛顿、埃及到黎巴嫩）购回的家具与装饰物件，以"旅游日志"的想法来重新为空间进行设计与装饰装修。

酒店空间采用绿色作为主色，将其呈现在多种层次色调与不同图案当中，搭配着紫红、暗灰、粉红、大红等颜色，再加上金色或是银色来收边。沙龙和电梯间的壁纸是在中国绘制的，沙发和座椅是英国制造的，一楼的桌子则是在贝鲁特购买的。设计师纳达－岱伯构思了一个由乌木和珍珠贝制成的吧台，吧台上方的玻璃隔板内镶着羽毛和亮片，与摆着彩色玻璃饰品的展示柜中的底衬银叶相呼应。酒店中亦使用着丝、绒、缎等珍贵的织品，例如大厅沙龙中的座椅和靠垫。

拥有世界风多样化菜单的餐厅融合了法国、地中海地区和亚洲的料理，完美地和这里的空间风格互相辉映，就像设计师塔法－莎兰说的："我希望让所有的旅客在停留巴黎的时间里都能够感到轻松自在，借助他们家乡文化当中的某些元素来让他们有一种像是在自己家里一样的感觉。"

No room is alike. The walls have been lined in toile de Jouy with Chinese motifs, chosen by the decorator ; the furniture is different each time: mirrors, chests and tables inset with mother-of-pearl ; bedheads and lamps in contemporary designs ; precious boxes and chests... The bathrooms are decorated with Moroccan zeliges or Italian marble ; with their glass doors, the cupboards resemble family heirlooms and the minibars are hidden behind wood carved in Syria.

这里没有两个房间是一模一样的，墙上裱着设计师精选的中国风织布，搭配不同的家具：镜子、橱柜、镶着珍珠贝的桌子、当代设计的床头柜和床头灯、收藏贵重物品的保险柜……。浴室里则采用了摩洛哥的瓷砖或是意大利的大理石来装修；带着玻璃门的衣柜像是传统的家族橱具。房间里的迷你吧台隐藏在叙利亚精制的原木板之后。

玛图汉酒店

Hôtel des Mathurins

Designer:	Vincent Bastie	设计师：	文森特-巴斯提
Location:	Paris, France	地点：	法国，巴黎
Completion Date:	2005	完工日期：	2005
Photographer:	Wijane Noree (pp.270, 272-277), Thomas Raffoux (p.171)	摄影师：	Wijane Noree (pp.270, 272-277), Thomas Raffoux (p.171)

The recently renovated standstone facade of the Hôtel des Mathurins is made up of a classical and elegant vocabulary. The entrance is signalled by a glass awning ornamented with lanterns. The hall, with its generous volumes, is surrounded by stucco walls and covered with a plaster ceiling with precious ornamentation. The marble floor is decorated with an original work by the artist de Rougemont, and the welcome desk and wall panelling are in natural oak.

The access to the floors above preserves the privacy of the occupants. A viewpoint opens up on the different lounges on the ground floor as well as towards the breakfast room. A lounge, decorated with acacia wood walls and brass ornaments, contains a library of art books devoted to Paris arranged around a chimneypiece...

On this floor there is also an intimate lounge bar where business guests can have a private meeting in a cosy atmosphere. The surprising curves of the staircase invite one to descend to the lower floor, which is lit by natural light from a generous glass roof situated just above. The spacious and airy breakfast room is sited away from the comings and goings of the hall.

On each floor the wide corridors are decorated in elegant striped fabrics with a different hue giving a recognisable identity to each floor. The rooms have a warm atmosphere thanks to fine fabrics covering the walls. A deep carpet covers the floor, the curtains are in silk velvet, the lighting is discreet and the latest technological equipment brings its own touches of comfort. The bathrooms are sober, light and well equipped. All these elements give the hotel a discreet and refined charm.

新近才重新整修完成的玛图汉酒店的石造立面呈现着古典优雅的建筑语汇，酒店入口因其上方装饰着灯笼的出挑棚罩而被彰显出来。宽敞的大厅四周围绕着由石泥涂漆的墙面，其上方有着精致的石膏浮雕天花板；地板大理石铺面上装饰着一件艺术家胡爵蒙的作品，接待处柜台以及墙面的饰板以原色橡木制成。

通往楼上的通道为住在酒店里的宾客维护了私密性，而在楼上的旅客则可以一览无遗地看见一楼的各个沙龙空间以及旅馆的早餐厅。其中一间以胡桃木装修墙面、带有铜锌合金装饰制品的沙龙里，有着一个大壁炉，壁炉附近摆设着一座专门收藏巴黎艺术图书的藏书架。

在相同的这个楼层，有一个较具有隐秘性的沙龙酒吧，让人能够在温暖宜人氛围中，安静地进行商务约会。此外，借着一座令人惊奇的曲线型楼梯通往地下层，则来到一个沐浴在由玻璃天顶引入自然天光的空间之中，这里是远离了大厅中来往人潮的早餐厅。

各客房楼层的宽大廊道都有个别独特的颜色作为识别，以优雅的条纹状织布来装修墙面。客房内部空间也采用了精致的织布来裱装其墙面，使得气氛更显亲切柔和；地上铺着厚厚的地毯，搭配着丝绒窗帘以及温和的照明，最新的科技设备更增加客房的舒适度；另外还有十分简洁、明亮、设备齐全的浴室。这所有的设计都让玛图汉酒店具有庄重雅致的魅力。

勒拉瓦锡酒店
Le Lavoisier Hotel

Designer:	Jean-Philippe Nuel	设计师：	让菲立浦-努埃勒
Location:	Paris, France	地点：	法国, 巴黎
Completion Date:	1999	完工日期：	1999
Photographer:	Le Lavoisier Hotel	摄影师：	Le Lavoisier Hotel

The hotel Le Lavoisier is situated in Paris' 8th arrondissement where the architecture was marked by the transformations of Baron Haussmann in the 19th century. At that time, Paris abandoned its still Medieval character to take on the face that we know today with its long avenues bordered with buildings which have remained as the city's identity.

The interior design of the hotel found its character in this history, and the decoration is the natural extension of the exterior architecture, marked by its classical vocabulary but revisited to make the hotel a contemporary project. The 30 rooms and suites of the hotel Le Lavoisier, divided over six floors, have been elegantly renovated with private terraces, furniture and antiques combined with beautiful paintings.

The hotel is perceived as a private interior, a Parisian apartment of today where the traveller has the impression of discovering what is hidden behind the facades of the grand avenues, in order to live and discover the interior city.

位于巴黎第八区的勒拉瓦锡酒店，十分具有19世纪欧斯曼男爵所推动的城市更新之后的建筑风格。在那个年代里，巴黎摆脱了中古世纪的城市面貌，一大部分转变成我们今日所熟知的、由笔直宽广大道与其两旁整齐建筑所组成的街区，成为现今巴黎的城市特征之一。

酒店的室内特色也受到这段历史的影响，它自然承袭了酒店外貌的古典建筑风格，但以全新的手法诠释古典，以符合当代精神。6层楼重新整建后的酒店共有30个一般客房和豪华套房，皆拥有私人阳台和家具、古物的装饰，并伴随着美丽的墙面色彩。

身在此地就像在巴黎某处的私人住宅中一般，让旅客们体验这个位于城市大道的建筑立面背后的空间，真正生活在城市的内部、发觉它的魅力。

加瓦尼酒店

Gavarni Hotel

Designer:	Françoise Morel
Location:	Paris, France
Completion Date:	2004
Photographer:	Xavier Moraga (pp.284, 285), Christophe Bielsa (pp.286, 287)

设计师：	弗朗索瓦－莫黑尔
地点：	法国，巴黎
完工日期：	2004
摄影师：	Xavier Moraga (pp.284, 285), Christophe Bielsa (pp.286, 287)

In the heart of Passy, the Gavarni Hotel offers a heaven of peace and refinement: 21 rooms and 4 suites are organised in a space with numerous possibilities... An alliance of luxury and modernity just a few steps from the Eiffel Tower.

In 2000, Françoise Morel was given the task of renovating the 2 upper floors, in which a suite, 3 junior suites and a deluxe room have been created. In 2004, the lobby, the restaurant rooms and a meeting room under a glass roof were entirely renovated, with the rooms and the corridors echoing this new decoration. Xavier Moraga, the director of the establishment, designed the furniture for the hotel: side tables in the bar, newspaper stands, sideboards for the breakfast buffet and the reception counter are a few examples.

The Gavarni Hotel combines the warmth and comfort of wood and the cheerful luminosity of limed paintwork. The decoration of each room is personalised, and each has a welcoming bathroom. Fine materials such as marble, oak and granite run throughout... The Eiffel, Trocadéro, Vendôme and Versailles suites enjoy a privileged view of the Eiffel Tower from their balconies, as well as the serenity of a living room area, a real little 18th-century boudoir with warm and refined tones that set off the antique furniture.

All the interior decoration was done in collaboration with Anne-Sophie Robert, an artist who specialises in patina and painted trompe-l'œil ceilings, which are found throughout, including in the new business centre. The plasterwork was done M. Hangar, a master craftsman of France. The works of Patrick Simasa and Anne-Sophie Robert, as well as numerous lithographs, adorn the establishment right through to the bedrooms.

位于巴黎市帕西区中心位置的加瓦尼酒店犹如一处宁静而雅致的避风港，它的21间标准客房以及4组套房，随着室内可多样变化的空间而配置。这个非常临近埃菲尔铁塔的酒店，是一个结合了豪华富丽以及现代感的空间。

在2000年，弗朗索瓦－莫黑尔负责为酒店最高两层楼里面重新设置一间豪华套房、三间头等套房以及一间豪华客房。在2004年，酒店的大厅、餐厅以及一个位于采光天顶之下的会议室也被重新整修，所有的客房以及走道空间都遵循了这个新的空间风格。酒店的总裁扎维埃－莫拉加甚至还亲自为它设计了家具，其中包括酒吧里的餐具桌、书报架、自助早餐的置餐台，以及接待处的柜台等。

加瓦尼酒店结合了木料的温暖舒适特性和粉刷墙面的明亮愉悦氛围，每一个房间都拥有个性化的装潢以及舒适宜人的浴室。设计师在此特意采用了像是大理石、橡木、花岗岩……等高贵的材料来塑造空间的性格，被命名为埃菲尔、托卡德侯、凡登以及凡尔赛的豪华套房都拥有面向埃菲尔铁塔景观的私人阳台。套房里小客厅的宁静氛围体现着18世纪贵族官邸女主人使用的隐秘沙龙的风格，此空间温暖细致的色调充分衬托出其中摆设之古董家具的价值感。

艺术家安苏菲－罗伯特参与了所有室内空间的设计，她擅长展现材料的长年色泽、古旧质感以及对天花板的处理，在整个酒店空间，甚至在其新开放的商务中心当中都可以看到她的创作。此外，由法国最佳的艺术工匠韩嘉所做的墙面石膏装饰也多处可见。其他几位艺术家…是派崔克史玛沙、安苏菲罗伯特的作品，以及为数众多的石板画，都一起装点着包括客房在内的各个酒店空间。

圣杰尔曼驿站酒店
Relais Saint-Germain

Designer:	Claudine Camdeborde
Location:	Paris, France
Completion Date:	Various
Photographer:	Denis Clément

设计师：	克劳汀-康德博尔德
地点：	法国，巴黎
完工日期：	定期翻修
摄影师：	Denis Clément

The 4 star hotel "Le Relais Saint-Germain" invites one to take a journey through time, history, art and culture. This hotel is a home away from home, where every single detail has been thought out in order to offer the highest level of comfort. The spacious rooms, the perfect location, the internationally acclaimed design and the helpful service together provide the perfect conditions in which to enjoy this Left Bank area to the full.

Guests enjoy a cosy and comfortable atmosphere where every room is a discovery in itself... All of them are spacious and each one is meticulously appointed in terms of its living space and its details, with the decoration incorporating a handsome arrangement of antiques and choice of fabrics. All these incredible rooms have their original wooden beams, precious materials and antique furniture...

The bedrooms offer very personalised spaces in which the traveller finds books, and armchairs inviting one to relax. The volumes of these rooms are all different, sometimes given atmosphere by exposed beams either on the ceilings, or dividing up the room into well defined spaces. Colours are both present and discreet, as if they knew when to warm the room up and when to hold back. The bathrooms with two-coloured marble decoration achieve a very welcoming kind of simplicity.

四星级的圣杰尔曼驿站酒店邀请人们在此度过一段充满历史、艺术与文化的时光之旅，酒店就像是旅客远离家园时候温暖的家，所有细节的注重都是为了提供最佳的舒适服务。宽敞的客房、完美的地点，加上充满设计感的空间和体贴入微的服务，圣杰尔曼驿站酒店是旅游者在巴黎塞纳河左岸的理想停留地点。

每一个房间对旅客而言都像是一个新发现，在舒适温柔的宽敞客房里，每一个生活空间与细节安排都经过精心设计。每个客房里都保有了建筑物原有的木梁，并使用高贵的材料以及古典家具，并注重古董和织品摆置装饰。

客房的空间的设计都十分个性化，旅客们在其中可以发现书籍与沙发，一个让人身心放松的角落。每个房间有着各不相同的空间体量，有的刻意强调外露的木梁结构，有的让天花板成为空间的重点，或是将客房界定成几个性格不同的区域。房间的色彩明显而婉约，时而较为耀眼，时而含蓄陪衬。浴室里的双色大理石处理让空间在简洁之中仍极具亲切感。

Escape - Relaxation - Zen
Dream Villas

度假别馆

自然–休闲–禅意

莎列特城堡酒店
Château de Salettes

Designer:	Cardete Huet Architectes, Gérard Tiné (plastic artist)
Location:	Cahuzac-sur-Vère, France
Completion Date:	1999
Photographer:	Gérard Cesar (pp.294-296, 297 top, 298-299),
	Christine Cayré (p.297 bottom)

设计师：	卡尔德特&宇越 联合建筑师事务所, 杰拉尔–提纳(造型艺术家)
地点：	法国, 卡韋扎克–苏维尔
完工日期：	1999
摄影师：	Gérard Cesar (pp.294-296, 297上图, 298-299),
	Christine Cayré (p.297下图)

Perched on its rock, the ruin of the Château de Salettes had a certain allure. Despite its decrepitude its silhouette still looked impressive over the undulations of the Gaillac vineyards. It seemed right to give the building the means to show its potential for spaces and volumes in order to perform a subtle metamorphosis: to bring out what modernity there was in the mass of the old stone built edifice in order to slide into the interior the comfort and refinement of contemporary living.

This sliding takes the form of successive envelopes which fold and interplay in the labyrinth of interior volumes. The envelopes are formed of many different materials which sometimes cover the stone and mask it, sometimes uncover it and expose its rusticity: an envelope of plaster that reframes the stone window openings ; a tactile envelope of wooden wainscoting ; a textile envelope that is soothing on the ear ; a white envelope formed of a film of lacquer that carries the visual dispersion of bursts of red, green, blue and yellow in the paleness of the rooms. Responding to the very simple and measured design of the successive envelopes is the design of the wooden furniture, whose volumes and surfaces fold and unfold in order to perform their functions.

It is neither nostalgia, nor a casual regard for the stones and for history. But the history of the place can continue from generation to generation with new ways of looking at things and new means of existing in the world.

莎列特城堡的废墟以其独特的姿态盘踞在它的岩石基础上, 尽管外表苍凉, 它的侧影仍然非常醒目地矗立在高低起伏的盖亚克区葡萄园山丘上。设计师善加运用了其空间与体量的特殊潜力, 在此进行了一个微妙的转型: 挖掘古老石造建筑的现代特质, 以便在室内空间内导入当代住所的舒适雅致。

在像迷宫一般的内部空间里, 展开了一序列的包裹装整的设计, 多样不同的材料时而覆盖着壁面的石材、时而将其外露以展现粗犷的质感, 例如: 用来框架石材门窗的石膏、触感极佳的木质护墙板、缓和听觉效果的织品布料、反射着红、绿、黄、蓝等光线的白色薄漆层。这些设计简洁、层层展开的裹覆材质, 搭配着可随不同需求而折叠伸缩的木质家具, 创造出各种不同的空间效果。

面对这里的历史痕迹与石材, 设计师并不特别赋予怀旧情思, 但也不忽视它的价值, 其所着重的是让这个场所的历史一代代相传下去, 让人们能以新的观点和方式来面对世界。

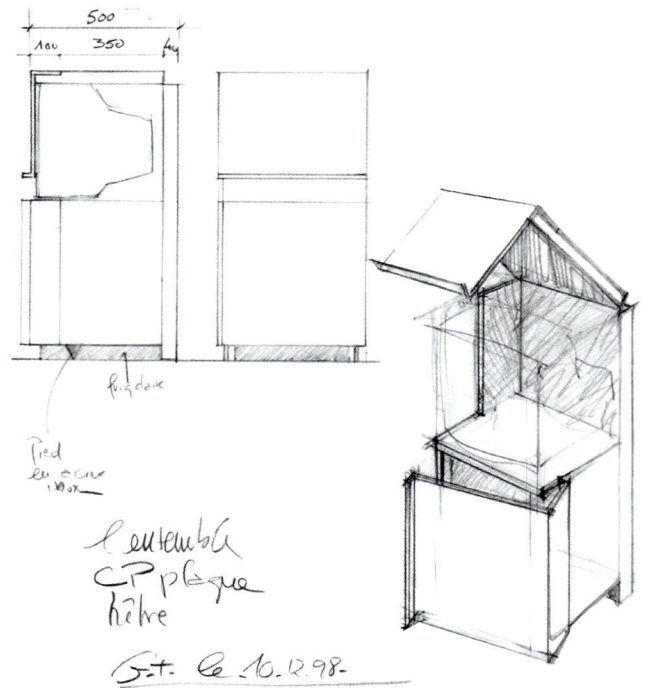

博家庄园酒店
Hameau des Baux

Designer:	Yann Hody, Phillippe Eckert
Location:	Le Paradou, France
Completion Date:	2004
Photographer:	Agence Caméléon

设计师:	伊安–侯迪, 菲立浦–艾克尔
地点:	法国, 帕哈度
完工日期:	2004
摄影师:	Agence Caméléon

Conceived as a family house, this property is a mix of ancestral traditions crossbred with a touch of contemporary daring. Refined and peaceful, the interior decoration restores all the splendour of Baux de Provence. Fine materials have been used to dress the restful spaces in a pure style. There are no bright colours, but instead muted, aged colours inspired by the history of the Alpilles.

Thought out with the purest regard for tradition, and with the love of old stone and materials of long ago, this hotel gives new life to the charming buildings of another age. A barn, a cottage, a chapel, a pigeon loft, a mill, decorated according to the theme of the room: a tabernacle door, a flight of pigeons, a stone trough or an "escoussin". All the materials and old stone were sourced from reclamation in order to give a special soul to the Hameau des Baux.

博家庄园酒店的建筑形态犹如一户家族住宅，并融合了古老的传统和一点当代的创新，其细致而平和的室内风格重新发扬了普罗旺斯地区博氏家族的辉煌荣耀。酒店内的休憩空间采用高贵的材质和纯净的设计风格，没有鲜艳的色彩，有的是呼应阿尔比勒山区历史的暗哑庄严的色调。

基于对大传统的尊重，以及对于古老石材与陈年质料的爱好，博家庄园酒店的设计为这些昔日建筑带来新生，仓库、小礼拜堂、鸽楼、磨坊等空间都一一按照不同主题来整修成为客房：圣柜之门、展翅之鸽、石造食槽……等。这里所有的古老材料或石头都是经过悉心搜寻采购而获得的，因而赋予了酒店一份特殊的灵气。

亚伯特一世庄园酒店
Hameau Albert 1ᵉʳ

Designer:	Bernard Ferrari
Location:	Chamonix, France
Completion Date:	2004
Photographer:	Nicolas Tosi (pp.310, 311, 313, 314), Philippe Schaff (pp.312, 315)

设计师：	贝尔纳-法拉利
地点：	法国, 霞慕尼
完工日期：	2004
摄影师：	Nicolas Tosi (pp.310, 311, 313, 314), Philippe Schaff (pp.312, 315)

The Hameau Albert 1er was originally a typical Chamonix hotel from the beginning of the 19th century. In the 1920s-30s it saw the construction of an Art Deco veranda. Extensions and additional floors followed in the 70s and 80s, and the building was reclad in a chalet style with brown weather-boarding and cut-out balconies. In 1993, a critical analysis of the whole was carried out by the architect Bernard Ferrari. A unison of beiges reaffirms the urbane, Belle Époque side of the building, replacing this "chalet" aspect. Inside, the rooms have gained in surface area. The dining rooms have also been redecorated by Bernard Ferrari. In Autumn 2000 an extension to the ground floor and first floor on the north facade was carried out, providing an adjoining kitchen and service area.

In 2004, the rooms of the main building were renovated, reducing them from 21 to 15. Their concept is simple: for each room, a print. Chamonix, being an important tourism and ski destination, has an exceptional collection of illustrations, which give each room its personality, its history. The subtle mauvy, pinky grey of the Aiguilles de Chamonix served as the main theme, from the curtains down to the stone used in the bathrooms. In these, the basins also evoke the world of Alpine peasants, with their pig trough form. The walls are in white and putty tones, the parquet in ash with under-floor heating. The linen rugs were designed and made to order, as well as the bed heads, writing desks and chests of drawers. Banquette-style beds and low aluminium tables were finished with a choice of fabrics and leathers that go appropriately with the pictures. The Japanese armchairs come in three different colours.

亚伯特一世庄园酒店最初极具有19世纪初期霞慕尼山区的旅店特色，在1920到1930年代的时候增建了一座新艺术风格的玻璃温室，此后酒店建筑不断扩建与加高，并在1970到1980年代间朝着山区木屋别墅的形式发展：运用棕色的原木外墙板、独立式阳台……等建筑语汇。在1993年，建筑师贝尔纳-法拉利受委托对酒店进行审慎的整体分析之后，决定以不同层次的灰褐色来整修其立面，以凸显其都市性格和代表法国两次大战之间美好年代的建筑风格，因而让酒店早期的山区木屋别墅形象完全改观。在室内空间方面，不仅客房的面积大为增加，贝尔纳-法拉利也对酒店餐厅重新进行设计装修。酒店更进一步在2000年的秋天为其地面层和北立面的二楼进行增建，以作为厨房与服务空间。

2004年，酒店为主楼客房进行重新装修，并将房数由21减低到15，新的设计概念十分简明：一个房间，一幅画。霞慕尼山区是著名的山间旅游胜地，具有十分出色而多样的景观，设计师利用它的景致为每一个房间塑造其独特的个性与故事。此山区特有的、带着一点紫红的微妙灰色成为设计发展的主轴，从窗帘到浴室的石材都可以发现它的踪迹。浴室的洗手台形似小猪的食槽，也让人联想到阿尔卑斯山的农村世界。室内灰白色的墙面搭配着经过热气处理的原木地板。此外，地面铺置的麻质地毯以及床头柜、托架式写字台、储物柜等家具都是特别为酒店设计与制作的。带有床头软垫的大床以及铝质的低矮桌子都经过精心摆设，并搭配着特选的布料与皮革，以和墙上的挂画和谐相衬。各房

The three farms that have been added to the Hamlet have been built using the old wood from the 15 old farms and chalets that were taken down. Bernard Ferrari was able to imagine a coherent whole, with each traditional element finding its own application. The technical side of the building, often invisible, is symbolised by the contemporary furniture and accessories combined with the treatment in old wood of most of the interior spaces.

依附着此庄园酒店的三个农庄建筑，是采用附近15个拆掉的老旧农舍或木屋别墅的回收木材所建造的，设计师贝尔纳–法拉利构想出一个整体和谐的配置方式，使得每一个回收的传统物件都在此找到它最适当的位置与用途。建筑的技术性通常不易直接显现，设计师在此则利用现代感十足的家具与配件来象征性地展现当代科技性格，运用它们来点画以旧木风格为主调的室内空间。

亚尔府邸酒店
Hôtel Particulier - Arles

Designer:	Brigitte Pagès de Oliveira		设计师：	碧姬-佩婕-德奥里维拉
Architect:	Paul Anouilh		建筑师：	保罗-阿努伊
Location:	Arles, France		地点：	法国，亚尔
Completion Date:	2006		完工日期：	2006
Photographer:	Bernard Touillon		摄影师：	Bernard Touillon

Five years separate the opening of the Hôtel Particulier and its enlargement project ; years during which Brigitte Pagès de Oliveira honed her eye, read, lived and traveled. From a legacy of several mixed tastes, the Hôtel Particulier shows the full extent to which a unique establishment can be created by putting the spirit back into an 18th-century building.

A first building, whose spirit is classicism revisited, punctuated by a few baroque touches, adjoins a newly renovated building. A second wing opens up with a more contemporary aesthetic language. Six rooms and suites have been brought to life with clear colours, where a tempo in black and white reveals a string of fine materials and carefully sought out objects: a black floor in quartz crystal and traditional blankets, leather or vinyl padded bedheads and a play of juxtaposed mirrors. Damask-upholstered Perspex chairs wink knowingly at the huge Murano chandelier, a self-confessed homage to Christian Lacroix.

Being a lover of the Arab world, Brigitte Pagès de Oliveira has introduced a light calligraphy in engraved glass to the hammam and bathrooms, together with basins designed by Philippe Starck. Then, unperturbed, she revisits Louis XV style for the chests of drawers which she combines with the comfortable Barcelona chairs of Mies van der Rohe.

From that moment on, the Hôtel Particulier gave birth to a world that offers a confidential promise of happiness to the fortunate few who are able to escape to the Camargue.

在酒店开幕和扩大经营相隔5年的时间里，酒店主人兼设计者的碧姬-佩婕-德奥里维拉透过阅读、生活和旅行的经历来让自己的鉴赏眼光更为犀利，以新颖的现代精神重新对这个18世纪建筑加以整治，使这个继承了多样品位的酒店展现出一种特殊的风貌。

酒店的第一个建筑体主要采用的是一种重新诠释过的古典主义风格，点缀着一些巴洛克式样的装饰。邻接此建筑体而新造的第二个建筑体则采用较为现代的美学语汇，其中6个房间与套房因采用了明亮的色调而显得生气盎然，黑白双色的搭配让装修空间的高贵材料和一些特别寻获的配件显得更为出色：黑色水晶石英地板、传统厚花布寝具、皮件或橡胶制成的软垫装修，以及制造影像叠加效果的镜面。碳纤维透光的座椅呼应着顶上穆拉诺制造的巨型水晶吊灯，仿佛向设计师克里斯汀-拉克鲁瓦致敬。

基于对阿拉伯世界的热爱，碧姬-佩婕-德奥里维拉还使用了浅写的毛笔字形来作为浴室以及蒸汽室玻璃上的刻印图案，与菲利浦-史塔克所设计的洗手台相映成趣。她并且大胆地运用新手法来诠释路易十五时代式样的橱柜，让它们与密斯-凡德罗所设计的巴塞隆纳椅搭配。

亚尔府邸酒店以它的环境质量与氛围悄悄保证了旅客在此的舒适愉悦，并且方便自此出游，在卡玛格地区度过美好时光。

K酒店
Hôtel K

Designer:	Bernard Wilhelm		设计师：	贝尔纳-威尔海姆
Location:	Baerenthal, France		地点：	法国, 巴洪达
Completion Date:	2006		完工日期：	2006
Photographer:	Frantisek Zvardon, Joern Stegen		摄影师：	Frantisek Zvardon, Joern Stegen

With a K as in Klein, the name of its creators Nicole and Jean-Georges, the fortunate hosts of the Arnsbourg restaurant in Baerenthal, the Hôtel K blends marvellously with the peerless forest charm of its chosen glade. A subtle mixture of wood and glass opens up to the exceptional setting of the Regional Park of Northern Vosges. With light walls, isolated low-voltage lighting, and the pure lines of its staircase, the Hôtel K is at once contemporary, sober and elegant.

Designed by the architect Bernard Wilhelm as a luxurious guest house, the K has a reception that feels just like a private home. The modern furniture is warmed up by the combination of wood and a fireplace that is always alight. A few Asian touches bring diversity and contrast to a setting where the range of colours and materials remains exclusively mineral and vegetal. The floors of the rooms are carpeted or in parquet, the walls have been painted in mineral tones, the lighting is adjustable to enable guests to choose an intimate or luminous atmosphere. These 12 rooms and suites play with the space and thus preserve transparency, notably in the bathrooms, which benefit from maximum luminosity. These bathrooms can be cordoned off or on the contrary open up onto the space thanks to translucent panels, making the space more intimate or creating a large loft.

On the walls, a few contemporary artworks bring colour and individuality to each of these living spaces, contrasting with the bedheads in solid wood or the driftwood lamps. Nature is a continual presence here. The "K", a work of architecture created from fluid lines and fine materials, flirts with art and makes this port of call a magical place where gastronomy goes hand in hand with tranquility and relaxation.

K就是克莱恩Klein的K，就是它的创建者妮科尔与让乔治的姓，他们两位也是巴洪达地方亚斯布尔餐厅的快乐主人。K酒店地处刻意精选过的森林中空地，并巧妙地利用了森林特有的魅力。其建筑结合了木材与玻璃两种材料，享受着面向法国北浮日地区国家自然公园的景观。酒店的室内有着明亮淡色的隔墙、点状配置的低电压照明、风格纯净的楼梯间，整体风格质朴雅致并且极具现代感。

以民宿的概念来建造，K酒店的接待厅让人仿佛身处一座私人宅院；由于多种木制材质的搭配组合，加上空间里有一座持续燃烧的壁炉，即使这里采用着极具现代感的家具，气氛仍然显得十分温暖。带点亚洲风的设计笔触，让室内空间里主要以矿岩和植物为基调的色彩与材质增加了多元性和对比感。房间地面铺着地毯或木质地板，墙面的粉刷则以岩石矿物色调为主，可以任意调节的照明让人在明亮舒畅或是幽暗私密两种情调中自由地选择。这共计12间的客房与套房都对空间格局有着精心构思，以保留日光的穿透性，特别是让浴室因此具有明亮的光线。借助着半透明的玻璃隔板，客房浴室可以根据不同情况需要而变化，选择向其他空间开放或是封闭独立以塑造私密感。

墙上的当代艺术作品为每一个空间带来不同的色彩与个性，并与木材制成的床头柜和灯具形成对比。充满自然风味的K酒店，本身就是精致的建筑作品，有着流畅的线条和高贵的材质，并散发着艺术的气息，为旅客的住宿停留，提供了一个宁静、轻松和具有美食享受的奇妙境地。

季瓦山公园酒店
Jiva Hill Park Hotel

Designer:	Jean-Philippe Nuel
Location:	Crozet, France
Completion Date:	2007
Photographer:	Fabrice Rambert

设计师：	让菲立浦-努埃勒
地点：	法国，克罗泽
完工日期：	2007
摄影师：	Fabrice Rambert

Ten minutes from Geneva airport, in a very natural setting, the Jiva Hill Park Hotel expresses the proximity of the mountains and the spirit of an international metropolis. It is the apparent opposition of these two worlds that gives the hotel its character and identity.

The lobby and the main gallery symbolise the hotel's values: an uncluttered, contemporary place that nevertheless uses fine and warm materials, such as slate, wood and different leathers. The restaurant and its bar are distinct from the hotel, with their own identity as a variation on the general decoration. Here the same modern spirit it's found with particular care taken to create an intimate atmosphere.

At Jiva Hill, the bedroom and the bathroom form a whole, a very luminous space that one can modify thanks to the large frosted glass panels. The bed is found in the middle of the space, facing the view, and the bathroom also benefits from a large opening on to the outside, with its bath and two dissymmetrical Corian sinks. The subtle gradation of colours brings a feeling of tranquility and softness.

Like a casket evoking the shade of an Asian forest with its mosaics and reflections, the Spa gives a touch of discreet luxury to a rather secret place.

距离日内瓦机场10分钟，地处大自然之中的季瓦山公园酒店，将就近的山景特色以及国际都会的精神同时展现出来，这两个截然不同的世界赋予了酒店独一无二的性格。

大厅和主要廊道呈现了酒店的主要设计概念：一个纯练优雅、足具当代感的场所，但运用的是板岩、木材、皮革等高贵温暖的材质来塑造空间。酒店别致的餐厅和吧台像是整体空间装修风格的一个变奏，拥有独特的性格，然而人们在此仍可以感受到和酒店其他空间相同的现代精神，以及同时塑造空间亲密气氛的刻意追寻。

酒店的客房是由卧室和浴室两者一起构成的明亮的整体性空间，因为采用可移动式半透明玻璃，使空间得以产生不同的组合效果。面对优美景观的卧床位于房间的中央，和浴室一样拥有对外开放的大面积开口；浴室里配备了浴缸和两座不对称的可丽耐材质的洗手台，加上彩色玉石的装饰使得空间更平静柔和。

这里的水疗健身中心(spa)犹如小珠宝盒让人联想到亚洲森林中光线明暗变化的效果，它全面运用了马赛克和水面映照的空间效果，让空间带有一种神秘、豪华婉约的气氛。

红杉庄园酒店
La Ferme du Domaine des Séquoias

Designer:	Pierre Buttion	设计师：	皮埃尔-卜雄
Location:	Bourgoin Jallieu-Ruy, France	地点：	法国, 布官-加柳-惠伊
Completion Date:	2006	完工日期：	2006
Photographer:	Jac Perrichon	摄影师：	Jac Perrichon

One can only conceive of inscribing one's existence in the future by marking one's one time with an authentic and modern imprint respectful of its heritage. This is the philosophy of the Domaine des Séquoias: respect for the past combined with a contemporary and original approach. Fourteen new rooms have been created in the farm adjoining the main residence where the restaurant is already found, to add to the five more classical rooms that were already there.

The main part of the restoration had to be done with respect for tradition. The framework was reconstructed using beams that were more than 150 years old, and the "scale" roofing tiles were replaced by new versions, hand-made and textured to match perfectly, the whole of the roof restoration being undertaken by the Compagnons du Devoir. So as not to betray the original function of this building, the rooms have been inserted seamlessly into the whole. The choice of architect was all-important: innovative and inventive, he or she had above all to be visionary.

The contemporary goes hand in hand with classicism and stone. The south facade formed of coloured glass panels in pink, yellow and blue, filters the rays of the sun and projects changing, coloured patterns into the interior. On the first floor, intrepid drops come to the fore in the form of vertiginous and transparent balconies suspended from the facade. Large picture windows extend along the stone walls. When it comes to the bathrooms, one can choose between a colourful bathroom with balneotherapy bath, a double bathtub, a single one or a wet room... Philippe Stark's MissK lamps bathe the rooms and their multicoloured furniture with their sophisticated rays, offering a lighting ambiance that can be altered to fit with any mood.

红杉庄园酒店将其对未来的展望奠基在对所处时代的认同，企图在其中留下一个名副其实、具有现代感并尊重过往的印记。它的哲学在于以一种尊重过去又十分具有当代创新精神的手法来进行经营。主要建筑设有5个古典客房与餐厅，其旁的农舍里面又新增了14间新的客房。

新房间的设置首先必须在尊重传统的原则下修复建筑物的结构体：重建的屋架系统采用了年龄超过150年的大木梁，鱼鳞状屋瓦是依照原样新作的，并且采用了经过手工上色的处理，整个屋顶由"德瓦伙伴"公司负责。为了符合建筑物作为酒店的最主要机能，客房单元的配置与传统风格大异其趣，建筑师在此对创新、革新与前瞻性的选择主导了整个设计的方向。

当代的设计风格因此和古典主义与古老石材共存，建筑物的南向立面，采用了粉红、黄、蓝三色的玻璃，让阳光经过过滤之后在室内产生多姿多彩的变化。在楼上有着大胆延伸的跳板、挑于立面之上而令人晕眩的透明阳台；同时，大面积的玻璃开窗分布在石造的墙面上。具有水疗设备、多彩的客房浴室里提供了多种选择搭配：双人浴盆、单人浴盆、意大利式淋浴间……。菲利浦-史塔克的MissK灯具点亮寝室，其光芒映照着彩色缤纷的家具，渲染出多种不同的明亮氛围。

波娃龙酒店
Le Beauvallon Hotel

Designer:	Olivier Gagnère	设计师：	奥利维耶–加尼尔
Location:	Sainte-Maxime, France	地点：	法国，圣马辛
Completion Date:	2007	完工日期：	2007
Photographer:	Franck Follet	摄影师：	Franck Follet

340+

Le Beauvallon has stood for almost a century above the gulf of Saint-Tropez. Its heart is equally balanced between Provence and the Mediterranean and has even ventured towards a touch of delicate orientalism since the renovation of the hotel in 1997 by its new Asian owners.

Built in 1913, le Beauvallon hotel is the work of the Swiss architect Julien Flegenheimer. Designed as a holiday establishment, it has had, since its beginnings, more than a hundred rooms and vast public spaces in a sumptuous Directoire style. The hotel is crossed from east to west by corridors serving the ground-floor salons and the rooms and suites over the four floors. On the ground floor, it is arranged around a rotunda that gives onto the terrace and the park.

In 1997, the interior decoration of the common spaces was entrusted to Olivier Gagnère. The reception spaces were adapted to the lifestyle of a 21st-century clientele. The ground floor abandoned its white walls and limed wood to be redecorated in subtle tones which are quite untypical for the Provençal setting. Jade green has invaded the South Salon and its superb rotunda. White Triton wall lights here shed a gentle light. They are also employed in the corridors, against the pearl-grey walls. Huge drapes in beige cotton and linen put the accent on the impressive height of the white ceilings and redistribute the volumes.

波娃龙酒店开立在圣特罗佩的海湾上有将近一个世纪的时间，原本兼具普罗旺斯与蔚蓝海岸的两种地方风格，在经过1997年由亚洲来的新东家重新整修后，更加添了些许的东方风情。

该酒店在1913年由瑞士建筑师朱里安–富来根海迈所建造，由于地处著名的度假胜地，酒店一开始便拥有上百间客房以及依照执政内阁时期(1795–1799)风格所设计的富丽堂皇的公共空间。旅馆内部由东西向的主要交通廊道来组织空间，包括地面层的数间沙龙，以及楼上4层的标准客房与豪华套房。在地面层的主轴廊道穿越一个面向公园和户外露台的中心圆厅。

酒店主人在1997年将公共空间的重整工程交给设计师奥利维耶–加尼尔，他所设计的接待空间完全针对21世纪旅客的需要而构思。在地面层的色彩规划上，他舍弃原有墙面的白色粉刷而改用了对于普罗旺斯地区而言较为罕见的淡彩色系。在南沙龙区及其所含括的圆厅则采用了翠绿色作为空间的主色，墙上的白色壁灯透射出柔和的光线，相同的壁灯延伸到走廊里，但衬底的墙面则使用了珍珠灰。棉麻混纺的羊毛色织布高挂在白色天花板下方，不仅强化了空间的高度，也重新分配了空间的体量。

In the fumoir, the wall lights are this time gilded with gold leaf. In the gastronomic restaurant "Les Colonnades", the warmth of crimson red awakens epicurean feelings and the characteristic line of the Marly armchairs prepares the gourmet comfortably for the flavours that will dance on his tastebuds.

The harmony of the volumes and respect for the authentic character of the place combine perfectly with a contemporary spirit and a few principles inspired by Feng Shui or scattered details of Asian culture. Le Beauvallon has thus been able to preserve the spirit of the eternal Riviera and its marvellous "art of living".

在吸烟厅里有着镶着金叶的壁灯，在名为"圆柱回廊"的美食餐厅里，胭脂红的热情氛围唤起人们享乐主义哲学所提倡的感官愉悦，马尔利名牌沙发的特殊线条让饕客们舒适地入座，准备以美食取悦其味蕾。

塑造整体空间体量的和谐感、对原有空间特性的尊重，这两个设计原则与一种具有当代感的空间精神、几个来自风水观念的装饰原则或者一些带有亚洲风格的细部处理完美地结合在一起，使今天的波娃龙酒店得以保存其永恒长河的精神以及其精致丰富的生活艺术。

凯里斯佩尔酒店
Le Lodge Kerisper

Designer:	Christophe Ducharme
Location:	La Trinité-sur-Mer, France
Completion Date:	2005
Photographer:	Guillaume Plisson

设计师：	克里斯多夫-杜查梅
地点：	法国, 特里尼泰港
完工日期：	2005
摄影师：	Guillaume Plisson

Le Lodge Kerisper, a charming hotel made up of 20 rooms, of which three are large suites, is situated beside the sea in the Breton region of Morbihan. It is a former collection of farm buildings transformed into a place for well-being, with an attractive swimming pool in a rectangle of teak, all organised around a terrace and interior garden planted with apple trees.

The interior decoration revives retro ambiances through different elements: parquet and antique mirrors, an old bar counter, and bistro tables for breakfast. Blanched wood and objects from flea markets contrast with designer lamps and touches of turquoise, almond or coral in the rooms. The bathrooms play with transparency behind "atelier" style windows: a subtle balance between old stone, painted wood, glass and metal.

凯里斯佩尔酒店是位于法国莫尔比昂省海边的魅力酒店，拥有20间客房，其中包括3间豪华套房。酒店的前身是个农庄，改建后成为一处提供人们舒解身心的场所，配备着一座镶在长方形木甲板中的游泳池，酒店的室内空间围绕着一个露天平台和内部的苹果树庭园来设置。

酒店的内部装修透过各种元素来重现复古风采：木地板、古镜面、老柜台、早餐厅里采用的旧时酒馆餐桌……等。在客房里，经过白色漆艺处理的木料搭配着四处搜购来的古物件，和带有强烈设计感的灯具以及杏仁绿、珊瑚红、土耳其绿等色彩形成鲜明的对比。浴室位于具有"工作室"风格的门窗之后，与卧室空间产生穿透性效果，并且将老石材、粉刷木、玻璃和金属等材料极为和谐细致地组合在一起。

马甘地乐酒店

Le Mas Candille

Designer:	Francis Chapus, Julie Avot, Brigitte Dumont de Chassart	设计师:	弗朗西斯-夏皮, 朱丽-阿沃, 碧姬特-杜蒙-德夏沙
Location:	Mougins, France	地点:	法国, 穆汕
Completion Date:	2006	完工日期:	2006
Photographer:	Laurent Loiseau	摄影师:	Laurent Loiseau

Situated in a park of 4.5 hectares, Le Mas Candille is surrounded by large cypresses and olive trees, and offers a superlative view. The earliest building dates from the 18th century. Le Mas Candille became a hotel in the 1960s. In 1999, a new owner transformed the place into a "small, charming, luxury hotel", which reopened in 2001 according to the plans of the architect Francis Chapus and the interior decorator Julie Avot.

Its individuality resides in an eclectic mixture of old style architecture and modern installations. Its rooms and suites, all different and spacious, are scattered over the carefully renovated Mas and the newly constructed Bastide. Some of them are extended by a terrace or a garden. Each piece of furniture, object and picture has a history ; they were found in antique shops and flea markets all over the world by Julie Avot.

The gastronomic restaurant "Le Candille" is very welcoming with dominant tones of green and red. Cut-out floral shapes on the wall expose stretched fabrics, allowing bronze on a green background and a play of transparency to appear... You would almost say that the vegetation had come inside the restaurant! Majestic red chandeliers make the place enchanting once night falls. A real cocoon in winter, it opens out onto the exterior during the summer period, in direct contact with nature.

And, designed by the architect Brigitte Dumont de Chassart, who was guided by her long experience in Asia, a Spa is found set back from the hotel, in the centre of a small Japanese garden.

马甘地乐酒店位于一个占地4.5公顷公园内，被柏树与橄榄树所环绕着，并拥有一望无际的视野。这里的第一栋建筑建于18世纪，在1960年以后成为酒店用途，在1999年，新的东主依照建筑师弗朗西斯-夏皮以及室内设计师朱丽-阿沃的方案，将这里转变成魅力豪华酒店，并且在2001年重新开幕。

这个酒店的独特性在于它巧妙地结合了古老风格与现代装置，形成一种折衷建筑风格。它间间相异、宽敞舒适的一般客房以及豪华套房，分布在整建后的马甘地乐酒店主楼和新建的巴斯提德馆。有些客房还拥有花园或阳台，让私人空间得以向外延伸。由室内设计师朱丽-阿沃从世界各地搜寻来的每一件家具、物件、画作背后都有一个特殊的故事。

马甘地乐酒店的美食餐厅以红绿双色为主色，形成活络热闹的氛围，加上有着花草叶文剪影的布料、绿底烘托着青铜，形成一种透明传参的游戏…，仿佛是外界的绿地深入到餐厅室内来。夜晚来临时，多盏红色壮丽的吊灯让餐厅变得魔幻迷人。餐厅在冬天有如温暖蚕茧，在风和日暖的季节则向户外开放，和大自然产生直接接触。

深处于一个小型日式庭园里的养生水疗中心(spa)，拥有远离酒店其他空间的静谧环境，是由具有丰富亚洲经验的建筑师碧姬特-杜蒙-德夏沙所设计。

两位修道院长酒店
Les Deux Abbesses Hotel

Designer:	Laurence Perceval Hermet (conception), Pierre Hermet (on site), Philippe Boudignon (safety architect)
Location:	Saint-Arcons d'Allier, France
Completion Date:	2005
Photographer:	Ferdinand Graf Von Luckner

设计师:	劳伦斯-柏士浮爱尔美(构思), 皮埃尔-爱尔美(特殊技术与工地协调), 菲立普-布迪农(建筑师-安全规范)
地点:	法国, 圣亚宫-阿利埃
完工日期:	2005
摄影师:	Ferdinand Graf Von Luckner

The lanes of the old village of Saint-Arcons d'Allier harbour an astonishing hotel, the symbol of a certain reconquest of the rural environment that intelligently combines safeguarding the built heritage and the creation of an economic activity as a source of employment.

Guests discover what is described as an "exploded" hotel. It is composed of a dozen rooms, or rather small houses restored in traditional style, which extend over the basalt village. The heart of the exploded hotel is the château, dating from the 12th century and restored in the 16th by the Marquis of Lafayette's pious ancestors, the abbesses Isabeau and Gabrielle de Lafayette. The hotel has been called Les Deux Abbesses to pay homage to them.

The entrace to the château hosts the reception, where a lacquered red dowry chest evokes accents of ancient China, taken up again humorously in the motif on the velvet of the sofas. The staircases of the Renaissance tower lead in a few steps to a dining room with manorial proportions, where the paintings of Alfons Alt portray a hare, the fetish of the house, in the manner of Dürer. By a listed spiral staircase in the tower, where the vestiges of a 16th-century fresco can be seen, one accedes to a vast blue salon where African furniture is mixed with art books.

在圣亚宫-阿利埃的老市区里有一个极为特殊的酒店，它是这里百年来沧海桑田、乡村转变为城镇的见证，并巧妙地将旧有建筑的保存和创造经济就业机会两者互相结合。

这是一个由12个客房组合而成的所谓"分散式"的酒店，其实是12个用传统方式整修过的小型的单栋建筑，散布在此地石造的小村庄里。酒店的核心单位位于一栋建于12世纪的城堡里，该城堡在16世纪由拉法叶男爵笃信宗教的祖先艺莎菠-拉法叶和加布里艾尔-拉法叶两位女修道院长重新修过，酒店的命名则是为了向她们致敬。

城堡的入口空间里设有接待处，婚用的大红衣橱令人联想到古时候的中国，同样的色彩也被诙谐地运用在绒布座椅的图案上。文艺复兴塔楼的阶梯带领人们前往尺度气派大方的餐厅，墙上挂着阿尔方斯-阿勒特以德国文艺复兴艺术家杜黑的绘画方式来描绘酒店吉祥物野兔的作品。塔楼里被列为保护的榫接楼梯间中有着16世纪留下来的壁画，人们可以借此楼梯通往一个宽敞的蓝色沙龙，其中摆设着来自非洲的家具和各种艺术书籍。

All around the château, following one another, are La Grange, a very dreamlike rural loft with its canopy supported by silver birch trunks ; La Maison Blanche, a vast organic residence on two levels, all curves and gentleness ; or La Cabane, a small house inspired by the vegetable and mineral with a surprising bathroom built into the eruptions of volcanic lava and a theatrical bedroom...

A former kitchen garden now houses an open-air swimming pool heated by the sun, which clings on to the southern slopes of the village and overlooks the Allier. Cheeky red "Nantucket style" benches invite one to pause for a moment in the sun.

城堡四周便是一系列的客房：名为"谷仓"的客房有着乡村式阁楼的梦幻感觉，其中有天盖的睡床是用桦树的树干所制成；名为"白宫"的客房是楼高两层的大别墅，有着圆润温和的线条；名为"棚屋"的客房空间较小，但却提供了由火山熔岩所雕塑成的惊奇浴室以及极具戏剧性的卧房，其设计灵感来自植栽和矿物的启发。

靠着村庄南面、居高临下俯视着阿利埃河的一块旧日菜园，如今成为利用阳光加温的露天游泳池，旁边散布着几张楠塔基特式的红椅凳，邀请人们在此和温驯的阳光一起歇息。

欧仁妮皇后温泉酒店
Les Prés d'Eugénie Hotel

Designer:	Michel et Christine Guérard	
Location:	Eugénie-les-Bains, France	
Photographer:	Michel Le Louarn (p.366), Laurent Parrault (pp.367, 369 top),	
	Tim Clinch (pp.368, 369 bottom, 370, 371 bottom, 373),	
	Guy Hervais (p.371 top), François Goudier (p.372)	

设计师：	米歇尔-盖哈尔, 克莉丝汀-盖哈尔
地点：	法国, 欧仁妮
摄影师：	Michel Le Louarn (p.366), Laurent Parrault (pp.367, 369 上图),
	Tim Clinch (pp.368, 369下图, 370, 371下图, 373),
	Guy Hervais (p.371上图), François Goudier (p.372)

In days gone by, the empress Eugénie loved this gorgeous colonial property combining several buildings in the shade of very beautiful gardens: the white Palais des Prés with its lacy balconies, the very pastoral Couvent des Herbes and the rustic Logis des Grives. Open to nature, a spacious and light succession of salons and dining rooms characterised by French style ceilings and baroque columns on pale stone floors awaits travellers.

In the shade of an ancient magnolia appears the Couvent des Herbes and its belfry. Damaged during the Revolution, cherished by Napoléon III, then brought back to life by Christine and Michel Guérard, it contains eight refined guest rooms: whitewashed walls and beams, waxed tomettes, soft rugs, canopy beds, fruitwood and antique 18th-and-19th century furniture, old master paintings, fireplaces. A haven of well-being where each room has his name and its soul. White dominates, awakened by the freshness of country fabrics. In this retreat, with its grey-blue timber framework, the sun floods in via two covered galleries.

An elegant rural house with dovecotes nestles in the meadow: the magical Ferme Thermale®, spa, where the virtues of thousands-of-years-old thermal springs lend themselves to those of medicinal plants, grown in the nearby garden.

该酒店建筑拥有大片花园、深具殖民地风格、曾经深受拿破仑三世欧仁妮皇后所钟爱。这里有带着雕栏阳台的白色"牧场别宫"、深具田园风味的"芳草修道院"、以及显现原野之气的"斑鸫之屋"；由连廊贯穿的几个沙龙与餐厅面向大自然而敞开，以宽敞而明亮的空间迎接着游人到来，其法兰西式的天花与立在黄石铺面上的巴洛克式柱为建筑带来优美韵律。

在百年玉兰树庇荫下的"芳草修道院"与它的排钟塔楼曾在法国大革命时期饱受摧残、之后获得拿破仑三世的珍视，而后又在克莉丝汀和米歇尔-圭拉尔两人的合力经营下重生。这里一共有8个精致的客房，有着石膏粉饰的墙面和梁柱、上蜡的陶土地砖、有天盖的床、果树木料制成的家具、加上18和19世纪风格的家具、古董画作和壁炉。"芳草修道院"犹如一处提供舒适安宁的避风港，每一个房间都有它自己的名字和灵魂，白色是这里的主色，搭配着田野风格、清新爽人的织布。这个世外桃源般的建筑有着灰蓝色的屋架，周围两道回廊的设计为室内带入充沛的阳光。

"温泉庄园"位于一栋隐藏在草原之中的优雅柱筋式建筑里面，是一个具有神奇性的水疗健康中心(spa)，以此地千年的温泉水结合了周边花园里出产的药用植物，来提供最佳疗效。

洛旭维兰酒店
Rochevilaine Hotel

Designer:	Anne Prohom		设计师:	安妮-普伦
Location:	Billiers, France		地点:	法国，毕耶
Completion Date:	2007		完工日期:	2007
Photographer:	Christian Vallée		摄影师:	Christian Vallée

On the promontory of Pen Lan, a hotel is "posed" on a rocky outcrop that sticks out between the ocean and the estuary of the Vilaine, from which its name of Rochevilaine comes. The beauty of the location is breathtaking: 300 metres of sea cliff plunging directly into the Atlantic... A monumental gate opens onto a unique collection of old buildings reconstructed stone by stone, the stunning whole having become a hotel with a unique appeal.

The 39 rooms and apartments are spread over country houses, a longhouse, a salt house, a granite castle and fishermen's houses by the water's edge... Each room has an individual character according to its architecture, its building materials, its decoration or its situation. Thus, the Manoir des Cardinaux houses a suite of two bedrooms, a living room and two terraces overlooking the ocean. The Chambre de l'Amiral is famous for its rotunda shape, from which several windows offer a 270° view of the ocean, allowing one to admire both the sunrise and sunset. The Loft des Artistes is ideal for lovers of design. Its 90 m² at water level and its 10 metres of opening onto the sea seem to "pull in" the light.

The feeling of being near the ocean floods the three restaurant rooms which, one by one, line up 47 metres of panoramic windows posed on the rocks. In 2007, a new contemporary decor was unveiled, creating a warm atmosphere with dark red arabesques, made dramatic by a play of lights.

The Domaine de Rochevilaine also houses a superb spa baptised "Aqua Phénicia" in memory of the Phoenicians who lived on this site in antiquity.

从容坐落于怪石峥嵘的磐澜海岬之上、向前突出位于大西洋和维兰湾之间的洛旭维兰酒店，因其所在的特殊地理位置而得名，这里有长达300米，面对大西洋令人屏息的美景。在经过一个宏伟的大门之后，抵达具有深厚历史而今日成为酒店的石造建筑物。

酒店有39个一般客房与公寓式套房，分别安置在小城堡、长条形库房、盐品海关、花岗岩城堡、水边的渔人住宅……等空间之中，每一个客房都因其所在的位置、建筑风格、材料而具有不同的个性。由主教小城楼所改建而成的套房里有两个房间、一个起居室以及两个面对大海的阳台；位在圆楼内的元帅客房，以其为数众多的窗户提供了270°的宽广视野，让人在此尽情观赏日出日落；而艺术家套房则是爱好设计人士的最爱，这个与海面齐平、面积90平方米的房间具有10米长的面海开窗，将光线大量引入室内。

位于岩石之上的酒店餐厅由3个不同的餐饮室组成，并有着长达47米的大片观景玻璃，仿佛漂浮在海洋之中。它在2007年呈现崭新的当代风貌，以曲线与光影交错相衬，沉浸在石榴红的温暖氛围之中。

在洛旭维兰酒店还有一个设备完善名为"腓尼基之水"的养生美颜中心，其命名是为了纪念古代在此地定居的腓尼基族人。

Hotel Directory

酒店资讯

3.14 Hotel
5 rue François Einesy
06400 Cannes, France
www.3-14hotel.com
info@hotel3-14.com
T +33 4 92 99 72 00
F +33 4 92 99 72 12

Balzac Hotel
6 rue Balzac
75008 Paris, France
www.hotelbalzac.com
reservation-balzac@jjwhotels.com
T +33 1 44 35 18 00
F +33 1 44 35 18 05

Bel Ami Hotel
7-11 rue Saint-Benoît
75006 Paris, France
www.hotel-bel-ami.com
contact@hotel-bel-ami.com
T +33 1 42 61 53 53
F +33 1 49 27 09 33

Bellechasse Hotel
8 rue de Bellechasse
75007 Paris, France
www.lebellechasse.com
info@lebellechasse.com
T +33 1 45 50 22 31
F +33 1 45 51 52 36

Cambon Hotel
3 rue Cambon
75001 Paris, France
www.cambon-hotel.com
contact@cambon-hotel.com
T +33 1 44 58 93 93
F +33 1 42 60 30 59

Cézanne Hotel
40 boulevard d'Alsace
06400 Cannes, France
www.hotel-cezanne.com
contact@hotel-cezanne.com
T +33 4 92 59 41 00

Chateaubriand Hotel
6 rue Chateaubriand
75008 Paris, France
www.hotelchateaubriand.com
welcome@hotelchateaubriand.com
T +33 1 40 76 00 50
F +33 1 40 76 09 22

Château Les Crayères
64 boulevard Henry Vasnier
51100 Reims, France
www.lescrayeres.com
reservation@lescrayeres.com
T +33 3 26 82 80 80
F +33 3 26 82 65 52

Château de Salettes
Lieu dit Salettes
81140 Cahuzac-sur-Vère, France
www.chateaudesalettes.com
salettes@chateaudesalettes.com
T +33 5 63 33 60 60
F +33 5 63 33 60 61

Daniel Hotel
8 rue Frédéric Bastiat
75008 Paris, France
www.hoteldanielparis.com
hotedanielparis@hoteldanielparis.com
T +33 1 42 56 17 00
F +33 1 42 56 17 01

Duo Hotel
11 rue du Temple
75004 Paris, France
www.parishotelleduo.com
contact@duoparis.com
T +33 1 42 72 72 22
F +33 1 42 72 03 53

Four Seasons - George V Paris
31 avenue George V
75008 Paris, France
www.fourseasons.com
T +33 1 49 52 70 00
F +33 1 49 52 70 10

François 1er Hotel
7 rue Magellan
75008 Paris, France
www.the-paris-hotel.com
hotel@hotel-francois1er.fr
T +33 1 47 23 44 04
F +33 1 47 23 93 43

Gavarni Hotel
5 rue Gavarni
75116 Paris, France
www.gavarni.com
reservation@gavarni.com
T +33 1 45 24 52 82
F +33 1 40 50 16 95

General Hotel
5-7 rue Rampon
75011 Paris, France
www.legeneralhotel.com
resa@legeneralhotel.com
T +33 1 47 00 41 57
F +33 1 47 00 21 56

Hameau Albert 1er
38 route du Bouchet
74402 Chamonix - Mont Blanc, France
www.hameaualbert.fr
infos@hameaualbert.fr
albert@relaischateaux.com
T +33 4 50 53 05 09
F +33 4 50 55 95 48

Hameau des Baux
Chemin de Bourgeac
13520 Le Paradou, France
www.hameaudesbaux.com
reservation@hameaudesbaux.com
T +33 4 90 54 10 30
F +33 4 90 54 45 30

Hôtel K
5 Untermuhlthal
57230 Baerenthal, France
www.arnsbourg.com
hotelk@wanadoo.fr
T +33 3 87 27 05 60
F +33 3 87 06 57 67

Hôtel des Mathurins
43 rue des Mathurins
75008 Paris, France
www.hotel-des-mathurins.com
contact@hotel-des-mathurins.com
T +33 1 44 94 20 94
F +33 1 44 94 00 44

Hôtel Particulier - Arles
4 rue de la Monnaie
13200 Arles, France
www.hotel-particulier.com
contact@hotel-particulier.com
T +33 4 90 52 51 40
F +33 4 90 96 16 70

Hôtel Particulier - Montmartre
23 avenue Junot
75018 Paris, France
www.hotel-particulier-montmartre.com
hotelparticulier@orange.fr
T +33 1 53 41 81 40

Jiva Hill Park Hotel
Route d'Harée
01170 Crozet, France
www.jivahill.com
welcome@jivahill.com
T +33 4 50 28 48 48
F +33 4 50 28 48 49

Keppler Hotel
10 rue Keppler
75016 Paris, France
www.hotelkeppler.com
hotel@keppler.fr
T +33 1 47 20 65 05
F +33 1 47 23 02 29

La Ferme du Domaine des Séquoias
54 Vie de Boussieu
38300 Bourgoin Jallieu-Ruy, France
www.domaine-des-sequoias.com
info@domaine-des-sequoias.com
T +33 4 74 93 78 00
F +33 4 74 28 60 90

Le Beauvallon Hotel
Boulevard des Collines
Beauvallon-Grimaud
83120 Sainte-Maxime, France
www.hotel-lebeauvallon.com
reservation@lebeauvallon.com
T +33 4 94 55 78 88
F +33 4 94 55 78 78

Le Bristol Hotel
112 rue du Faubourg Saint-Honoré
75008 Paris, France
www.hotel-bristol.com
resa@lebristolparis.com
T +33 1 53 43 43 00
F +33 1 53 43 43 01

L'Empire Hotel
48 rue de l'Arbre Sec
75001 Paris, France
www.lempire-paris.com
contact@lempire-paris.com
T +33 1 40 15 06 06
F +33 1 40 15 06 70

Le Grand Hôtel
12 place de la Gare
67000 Strasbourg, France
www.hotelspreference.com/strasbourg
strasbourg@hotelspreference.com
T +33 3 88 52 84 84

Le Lavoisier Hotel

21 rue Lavoisier
75008 Paris, France
www.paris-hotel-lavoisier.com
info@hotellavoisier.com
T +33 1 53 30 06 06
F +33 1 53 30 23 00

Le Lodge Kerisper

4 rue du Latz
56470 La Trinité-sur-Mer, France
www.lodgekerisper.com
contact@lodgekerisper.com
T +33 2 97 52 88 56
F +33 2 97 52 76 39

Le Mas Candille

Boulevard Clément Rebuffel
06250 Mougins, France
www.lemascandille.com
info@lemascandille.com
T +33 4 92 28 43 43
F +33 4 92 28 43 40

Le Méridien - Paris Étoile

81 boulevard Gouvion Saint-Cyr
75017 Paris, France
www.starwoodhotels.com
guest.etoile@lemeridien.com
T +33 1 40 68 34 34
F +33 1 40 68 31 31

Le Petit Moulin Hotel

29-31 rue du Poitou
75003 Paris, France
www.paris-hotel-petitmoulin.com
contact@hoteldupetitmoulin.com
T +33 1 42 74 10 10
F +33 1 42 74 10 97

Le Quartier Bastille Hotel

9 rue de Reuilly
75012 Paris, France
www.lequartierhotelbf.com
resa@lequartierhotelbf.com
T +33 1 43 70 04 04
F +33 1 43 70 96 53

Le Rex Hotel

10 cours Gambetta
65000 Tarbes, France
www.lerexhotel.com
reception@lerexhotel.com
T +33 5 62 54 44 44
F +33 5 62 54 45 45

Les Ateliers de l'Image

36 boulevard Victor Hugo
13210 Saint-Rémy-de-Provence, France
www.hotelphoto.com
info@hotelphoto.com
T +33 4 90 92 51 50
F +33 4 90 92 43 52

Les Deux Abbesses Hotel

Le Château, Le Bourg
43300 Saint-Arcons d'Allier, France
www.lesdeuxabbesses.com
abbesses@relaischateaux.com
T +33 4 71 74 03 08
F +33 4 71 74 05 30

Les Prés d'Eugénie Hotel

40320 Eugénie-les-Bains, France
www.michelguerard.com
reservation@michelguerard.com
T +33 5 58 05 05 05
F +33 5 58 51 10 10

Le Vignon Hotel

23 rue Vignon
75008 Paris, France
www.levignon.com
reservation@hotelvignon.com
T +33 1 47 42 93 00
F +33 1 47 42 04 60

L'Hôtel

13 rue des Beaux-Arts
75006 Paris, France
www.l-hotel.com
stay@l-hotel.com
T +33 1 44 41 99 00
F +33 1 43 25 64 81

Maison Pic

285 avenue Victor Hugo
26000 Valence, France
www.pic-valence.com
office@pic-valence.com
T +33 4 75 44 15 32
F +33 4 75 40 96 03

Marignan Champs-Élysées Hotel

12 rue de Marignan
75008 Paris, France
www.hotelmarignan.fr
reservation@hotelmarignan.fr
T +33 1 40 76 34 56
F +33 1 40 76 34 34

Montalembert Hotel

3 rue de Montalembert
75007 Paris, France
www.montalembert.com
concierge@montalembert.com
T +33 1 45 49 68 68
F +33 1 45 49 69 49

New Hotel of Marseille

71 boulevard Charles Livon
13007 Marseille, France
www.newhotelofmarseille.com
info@newhotelofmarseille.com
T +33 4 91 31 53 15
F +33 4 91 31 20 00

Park Hyatt – Paris Vendôme

5 rue de la Paix
75002 Paris, France
www.paris.vendome.hyatt.fr
vendome@hyattintl.com
T +33 1 58 71 12 34
F +33 1 58 71 12 35

Pershing Hall Hotel

49 rue Pierre Charron
75008 Paris, France
www.pershinghall.com
info@pershinghall.com
T +33 1 58 36 58 00
F +33 1 58 36 58 01

Plaza Athénée Hotel

25 avenue Montaigne
75008 Paris, France
www.plaza-athenee-paris.fr
T +33 1 53 67 66 65
F +33 1 53 67 66 66

Relais Saint-Germain

9 carrefour de l'Odéon
75006 Paris, France
www.hotelrsg.com
hotelrsg@wanadoo.fr
T +33 1 43 29 12 05
F +33 1 46 33 45 30

Rochevilaine Hotel

Pointe de Pen Lan
56190 Billiers, France
www.domainerochevilaine.com
domaine@domainerochevilaine.com
T +33 2 97 41 61 61
F +33 2 97 41 44 85

Scribe Hotel

1 rue Scribe
75009 Paris, France
www.sofitel.com
h0663@accor.com
T +33 1 44 71 24 24
F +33 1 42 65 39 97

Sheraton Paris Airport

CDG Airport
BP 35051 Tremblay-en-France, France
95716 Roissy Cedex
www.sheratonparisairport.fr
T +33 1 49 19 70 70

The Five Hotel

3 rue Flatters
75005 Paris, France
www.thefivehotel.com
contact@thefivehotel.com
T +33 1 43 31 74 21

The Westin Paris

3 rue de Castiglione
75001 Paris, France
www.westin.com
reservation.01729@westin.com
T +33 1 44 77 11 11
F +33 1 44 77 14 60

Troisgros Hotel

Place Jean Troisgros
42300 Roanne, France
www.troisgros.fr
info@troisgros.com
T +33 4 77 71 66 97
F +33 4 77 70 39 77

Architect & Designer Directory

设计师资讯

Pascal Allaman

12 rue Villehardouin
75003 Paris, France
T +33 1 48 24 06 57
pascalallaman@free.fr

pp.08-19, Bel Ami Hotel

Paul Andreu

15 rue du Parc Montsouris
75014 Paris, France
T +33 1 58 10 05 15
F +33 1 53 62 02 20
www.paul-andreu.com
carole.rami@paul-andreu.com

pp.148-153, Sheraton - Paris Airport

Paul Anouilh

20 avenue de la Libération
13200 Arles, France
T +33 4 90 49 66 96
F +33 4 90 93 94 57
www.anouilh-architecte-provence.com
paul.anouilh@wanadoo.fr

pp.316-321, Hôtel Particulier - Arles

Atelier Cardete & Huet

38 rue Alfred Duméril
31400 Toulouse, France
T +33 5 61 53 76 02
F +33 5 61 25 99 42
www.cardete-huet.com
agence@ch.tm.fr

pp.294-299, Château de Salettes

Julie Avot

Altodesign
Résidence Le Grand Hôtel
45 boulevard de la Croisette
06400 Cannes, France
T +33 4 93 68 71 93
F +33 4 93 68 74 33
M +33 6 60 71 47 56
julie@altodesign.fr

pp.354-359, Le Mas Candille

Vincent Bastie

6 rue du Parc
94160 Saint-Mandé, France
T +33 1 43 74 10 06
F +33 1 41 74 06 01
cabinetbastie.vincent@hotmail.fr

pp.26-37, Le Petit Moulin Hotel
pp.48-53, The Five Hotel
pp.270-277, Hôtel des Mathurins

**Nathalie Battesti &
Véronique Terreaux**

28 rue Beaurepaire
75010 Paris, France
T +33 1 42 02 36 28
F +33 1 42 02 36 98
www.ideogalerie.com
info@ideogalerie.com

pp.08-19, Bel Ami Hotel

Isabelle Bonis

Semi Tarbes
29 rue Georges Clémenceau
65000 Tarbes, France
T +33 5 62 51 78 51
F +33 5 62 44 16 93

pp.38-43, Le Rex Hotel

Philippe Boudignon

Scpa Allibert Boudignon
21 avenue des Belges
43000 Le Puy en Velay, France
T +33 4 71 02 01 21
F +33 4 71 02 11 38
www.ab-archis.fr
ab43@tiscali.fr

pp.360-365, Les Deux Abbesses Hotel

Pierre Buttion

Switch Architectes
24 rue Godefroy
69006 Lyon, France
T +33 4 37 48 88 90
F +33 4 37 48 88 91
www.archiswitch.com
info@archiswitch.com

pp.336-339, La Ferme du Domaine des Séquoias

Francis Chapus

Atelier Francis Chapus - Architectes Urbanistes
455 Promenade des Anglais
Le Quadra
06200 Nice Cedex 3, France
T +33 4 93 21 05 15
F +33 4 93 21 05 22
www.architectes-cote-azur.com
atelier.chapus@wanadoo.fr

pp.354-359, Le Mas Candille

Claudine Camdeborde

Relais Saint-Germain Hotel
9 carrefour de l'Odéon
75006 Paris, France
T +33 1 44 27 07 97
F +33 1 46 33 45 30
www.hotelrsg.com
hotelrsg@wanadoo.fr

pp.288-291, Relais Saint-Germain

Philippe Daraux

27 rue du Cherche Midi
75006 Paris, France
T +33 1 45 48 83 79
F +33 1 45 44 21 11
www.misendemeure.com
philippe.daraux@misendemeure.com

pp.250-255, Chateaubriand Hotel

Pascale Douillard

Agence Parallelom
5 rue du Bray
78400 Chatou, France
T +33 6 70 27 33 75
pa.douillard@wanadoo.fr

pp.154-157, Le Quartier Bastille Hotel

Joris Ducastaing

Atelier d'architecture
9 place Verdun
65000 Tarbes, France
T +33 5 62 37 11 97
F +33 5 62 51 22 52
www.ducastaing.fr
contact@ducastaing.fr

pp.38-43, Le Rex Hotel

Christophe Ducharme

Villa des Arts
15 rue Hégésippe Moreau
75018 Paris, France
T +33 1 45 22 07 75
F +33 1 45 22 07 76
c.ducharme.architecte@wanadoo.fr

pp.346-353, Le Lodge Kerisper

Brigitte Dumont de Chassart

Latitudes
10 rue des Filles du Calvaire
75003 Paris, France
T +33 1 40 27 90 87
F +33 1 44 61 85 00
latitudes.consultants@wanadoo.fr

pp.354-359, Le Mas Candille

Philippe Eckert

Mas de la Dame
529 bis chemin Saint-Gabriel
13160 Chateaurenard, France
T +33 4 32 62 10 45
P +33 6 09 20 29 64
F +33 4 32 62 19 52
philippeeckert@wanadoo.fr

pp.300-309, Hameau des Baux

**Alexandra Ellena &
Karine Ellena-Partouche**

T +33 4 92 99 72 00
F +33 4 92 99 72 12
aleroy@g-partouche.fr

pp.60-67, 3.14 Hotel

Jean-François Force

Cabinet Clé Millet International
21 rue de Bièvre
75005 Paris, France
T +33 1 53 10 11 66
F +33 1 53 10 11 67
www.clemilletinternational.com
cle.millet@noos.fr

pp.98-105, Cambon Hotel

Bernard Ferrari

La Griaz
74310 Les Houches, France
T +33 4 50 54 43 93
F +33 4 50 54 46 01
bernard.ferrari@wanadoo.fr

pp.310-315, Hameau Albert 1er

Olivier Gagnère & Associés

47 boulevard Saint-Jacques
75014 Paris, France
T/F +33 1 45 80 79 67
www.gagnere.net
olivier@gagnere.net

pp.256-261, Marignan Champs-Élysées Hotel
pp.340-345, Le Beauvallon Hotel

Jacques Garcia

212 rue de Rivoli
75001 Paris, France
T +33 1 42 97 48 70
F +33 1 42 97 48 73
www.decojacquesgarcia.com
cwhite@decojacquesgarcia.com

pp.188-193, The Westin Paris
pp.234-241, L'Hôtel

Jacques Grange

Jacques Grange Office
118 rue du Faubourg Saint-Honoré
75008 Paris, France
T +33 1 47 42 49 39
T +33 1 42 66 24 17
contact@jacquesgrange.com

pp.182-187, Scribe Hotel

Michel et Christine Guérard

Les Prés d'Eugénie Hotel
40320 Eugénie-les-Bains, France
T +33 5 58 05 05 05
F +33 5 58 51 10 10
www.michelguerard.com

pp.366-373, Les Prés d'Eugénie Hotel

Yann Hody et Christine Cauquil

126 avenue de Stalingrad
13200 Arles, France
T +33 4 90 93 70 44
F +33 4 90 93 16 53
www.cauquil-hody.fr
contact@cauquil-hody.fr

pp.300-309, Hameau des Baux

Patrick Jouin

8 passage de la Bonne Graine
75011 Paris, France
T +33 1 55 28 89 20
F +33 1 58 30 60 70
www.patrickjouin.com
agence@patrickjouin.com

p.199, bar in Plaza Athénée Hotel

Christian Lacroix

73 rue du Faubourg Saint-Honoré
75008 Paris, France
T +33 1 42 68 79 00
www.christian-lacroix.fr
info@c-lacroix.com

pp.26-37, Le Petit Moulin Hotel
pp.54-59, Bellechasse Hotel

Roland De Leu

26 rue des Plantes
75014 Paris, France
T +33 1 45 41 26 25
F +33 1 45 41 40 67
de-leu@wanadoo.fr

pp.114-119, L'Empire Hotel

Christian Liaigre

61 rue de Varenne
75007 Paris, France
T +33 1 47 53 78 76
www.christian-liaigre.fr
sales@christian-liaigre.fr

pp.176-181, Montalembert Hotel

**Laurence Perceval Hermet &
Pierre Hermet**

Hôtel Eclaté Les Deux Abbesses
Le Château – Le Bourg
43300 Saint-Arcons d'Allier, France
T +33 4 71 74 03 08
F +33 4 71 74 05 30
www.lesdeuxabbesses.com
abbesses@relaischateaux.com

pp.360-365, Les Deux Abbesses Hotel

Sybille de Margerie
9 rue Emile Allez
75017 Paris, France
T +33 1 40 55 70 70
F +33 1 40 55 70 71
www.smdesign.fr
smdesign@smdesign.fr

pp.188-193, The Westin Paris

Jacques Mechali
42 rue du Docteur Roux
75015 Paris, France
T +33 1 47 83 48 62
F +33 1 42 73 34 19
jacquesmechali@aol.com

pp.138-141, Le Vignon Hotel

Jacques Molho
11 rue de Belfort
67000 Strasbourg, France
T +33 3 88 41 05 34
F +33 3 88 84 48 04
Molho.jacques@wanadoo.fr

pp.110-113, Le Grand Hôtel

Jean-Philippe Nuel
9 boulevard de la Marne
94130 Nogent-sur-Marne, France
T +33 1 45 14 12 10
F +33 1 48 77 26 92
www.jeanphilippenuel.com
jpn@jeanphilippenuel.com

pp.84-89, Duo Hotel
pp.106-109, General Hotel
pp.278-283, Le Lavoisier Hotel
pp.330-335, Jiva Hill Park Hotel

Brigitte Pagès de Oliveira

L'Hôtel Particulier – Arles
4 rue de la Monnaie
13200 Arles, France
T +33 4 90 52 51 40
F +33 4 90 96 16 70
www.hotel-particulier.com
contact@hotel-particulier.com

pp. 316-321, Hôtel Particulier – Arles

Mathieu Paillard

Agent M
20 rue Moreau
75012 Paris, France
T +33 1 44 73 41 19
F +33 1 40 19 06 91
www.agentm.fr
agentm@agentm.fr

pp.20-25, Hôtel Particulier – Montmartre

Roland Paillat

Le Village
04300 Dauphin, France
T +33 4 92 79 56 41
r.paillat@wanadoo.fr

pp.158-163, Les Ateliers de l'Image

Anne Prohom

Acanthe Décoration
17 rue Emile Dubois
75014 Paris, France
T +33 1 45 88 79 19
P +33 6 03 05 01 51
www. rouge-et-or.com
prohom.anne@orange.fr

pp.374-379, Rochevilaine Hotel

Andrée Putman

83 avenue Denfert-Rochereau
75014 Paris, France
T +33 1 55 42 88 55
T +33 1 55 42 88 50
www.andreeputman.com
archi@andreeputman.com

pp.44-47, Pershing Hall Hotel
pp.148-153, Sheraton – Paris Airport

Philippe Puvieux

331 Corniche Architectes
331 Corniche Kennedy
13007 Marseille, France
T +33 4 96 20 31 10
F +33 4 96 20 31 11
www.331corniche-architectes.com
331.corniche.architects@wanadoo.fr

pp.68-75, Maison Pic

Imaad Rahmouni

8 passage de la bonne graine
75011 Paris, France
T +33 1 40 21 01 05
F +33 1 40 21 01 27
www.imaadrahmouni.com
contact@imaadrahmouni.com

pp.44-47, Pershing Hall Hotel

Pierre-Yves Rochon

9 avenue Matignon
75008 Paris, France
T +33 1 44 95 84 84
F +33 1 44 95 84 70
www.pyr-design.com
info@pyr-design.com

pp.120-127, Le Méridien – Paris Étoile
pp.128-137, Keppler Hotel
pp.188-193, The Westin Paris
pp.222-227, Château Les Crayères
pp.228-233, Four Seasons – George V Paris
pp.244-249, François 1er Hotel

Tarfa Salam
Tarfa Salam Design
6 Addisland Court
Holland Villas Road
London W14 8DA, UK
T +44 20 7603 1089
F +44 20 7610 5172
tsalam@tarfasalamdesign.com

pp.262-269, Daniel Hotel

Alain Sarles

Archimed
79 rue Liandier
13008 Marseille, France
T +33 4 91 80 26 26
F + 33 4 91 80 45 92
www.archimed13.com
archimed.1@archimed13.com

pp.76-83, New Hotel of Marseille

Elodie Sire

Schmidt et Mamann Design
29 rue d'Artois
75008 Paris, France
T +33 1 45 63 72 72
F +33 1 45 63 49 59
esire@schmidt-mamann-design.com

pp.250-255, Chateaubriand Hotel

Gérard Tiné

Domaine du Fréau
31180 La Peyrouse-Fossat, France
T +33 5 61 09 18 17
P +33 6 84 31 67 15
g.tine@free.fr

pp.294-299, Château de Salettes

Ed Tuttle

Ed Tuttle – Design Realization
71 rue des Saints Pères
75006 Paris, France
T +33 1 42 22 65 77
F +33 1 42 22 81 37
designrealization@designrealization.net

pp.142-147, Park Hyatt – Paris Vendôme

Bernard Wilhelm

2 rue du Nord
67800 Bischheim, France
T +33 3 88 81 97 44
F +33 3 88 81 86 95
bernardwilhelm@aol.com

pp.322-329, Hôtel K

图书在版编目（CIP）数据

　　法国酒店设计 / 法国亦西文化编 ；林明炘译 . — 沈
阳 ：辽宁科学技术出版社，2017.6
　　ISBN 978-7-5591-0132-7

　　Ⅰ．①法… Ⅱ．①法… ②林… Ⅲ．①饭店－建筑
设计－法国－图集 Ⅳ．① TU247.4-64

　　中国版本图书馆 CIP 数据核字（2017）第 072593 号

出版发行：辽宁科学技术出版社
　　　　　（地址：沈阳市和平区十一纬路 25 号　邮编：110003）
印　刷　者：辽宁新华印务有限公司
经　销　者：各地新华书店
幅面尺寸：240mm×320mm
印　　张：48
插　　页：4
字　　数：150 千字
出版时间：2017 年 6 月第 1 版
印刷时间：2017 年 6 月第 1 次印刷
责任编辑：杜丙旭 韩欣桐
封面设计：何　萍
版式设计：何　萍
责任校对：东　戈

书　　　号：ISBN 978-7-5591-0132-7
定　　　价：358.00 元

编辑电话：024-23280035
邮购热线：024-23284502
http://www.lnkj.com.cn